Lecture Notes in Mathematics 1569

Editors:
A. Dold, Heidelberg
B. Eckmann, Zürich
F. Takens, Groningen

Vilmos Totik

Weighted Approximation
with Varying Weight

Springer-Verlag

Berlin Heidelberg New York
London Paris Tokyo
Hong Kong Barcelona
Budapest

Author

Vilmos Totik
Bolyai Institute
University of Szeged
Aradi v. tere 1
6720 Szeged, Hungary
and
Department of Mathematics
University of South Florida
Tampa, FL 33620, USA

Mathematics Subject Classification (1991): 41A10, 41A17, 41A25, 26Cxx, 31A10, 31A99, 41A21, 41A44, 42C05, 45E05

ISBN 3-540-57705-X Springer-Verlag Berlin Heidelberg New York
ISBN 0-387-57705-X Springer-Verlag New York Berlin Heidelberg

Library of Congress Cataloging-in-Publication Data. Totik, V. Weighted approximation with varying weights / Vilmos Totik. p. cm. – (Lecture notes in mathematics; 1569) Includes bibliographical references and index. ISBN 3-540-57705-X (Berlin: softcover: acid-free). – ISBN 0-387-57705-X (New York: acid-free)
1. Approximation theory. 2. Polynomials. I. Title. II. Series: Lecture notes in mathematics (Springer-Verlag); 1569. QA3.L28 no. 1569 [QA221] 510 s–dc20 [511'.42] 93-49416

© Springer-Verlag Berlin Heidelberg 1994
Printed in Germany

SPIN: 10078788   46/3140-543210 - Printed on acid-free paper

Contents

1   Introduction

In this work we are going to discuss polynomial approximation with weighted polynomials of the form $w^n P_n$, where w is some fixed weight and the degree of P_n is at most n. We emphasize that the exponent of the weight w^n changes with n, so this is a different (and in some sense more difficult) type of approximation than what is usually called weighted approximation. In fact, in the present case the polynomial P_n must balance exponential oscillations in w^n. To have a basis for discussion let us consider first an important special case.

Let $w(x) = \exp(-c|x|^\alpha)$, $c > 0$ be a so called Freud weight. H. N. Mhaskar and E. B. Saff [34] considered weighted polynomials of the form $w^n P_n$, where the degree of P_n is at most n. They found that the norm of these weighted polynomials live on a compact set S_w, i.e. for every such weighted polynomial we have

$$\|w^n P_n\|_{\mathbf{R}} = \|w^n P_n\|_{S_w},$$

furthermore, $w^n P_n / \|w^n P_n\|_{S_w}$ tends to zero outside S_w. They also explicitly determined S_w:

(1.1) $$S_w = [-\gamma_\alpha^{1/\alpha} c^{-1/\alpha}, \gamma_\alpha^{1/\alpha} c^{-1/\alpha}]$$

where

$$\gamma_\alpha := \int_0^1 \frac{v^{\alpha-1}}{\sqrt{1-v^2}}\, dv = \Gamma(\frac{\alpha}{2})\Gamma(\frac{1}{2})/(2\Gamma(\frac{\alpha}{2} + \frac{1}{2})),$$

(see Section 3 below).

One of the most challenging problems of the eighties in the theory of orthogonal polynomials was Freud's conjecture (see Section 3) about the asymptotic behavior of the recurrence coefficients for orthogonal polynomials with respect to the weights w. The solution came in three papers [16], [29] and [27] by D. S. Lubinsky, A. Knopfmacher, P. Nevai, S. N. Mhaskar and E. B. Saff. The most difficult part of the proof was the following approximation theorem ([29]).

Theorem 1.1 *If $w_\alpha(t) = \exp(-\gamma_\alpha|t|^\alpha)$, $\alpha > 1$ is a Freud weight normalized so that $S_{w_\alpha} = [-1, 1]$, then for every continuous f which vanishes outside $(-1, 1)$ there are polynomials P_n of degree at most n, $n = 1, 2, \ldots$ such that $w_\alpha^n P_n$ uniformly tends to f on the whole real line.*

Let us mention that it follows from what we have said about $w_\alpha^n P_n$ tending to zero outside $[-1, 1]$, if f can be uniformly approximated by $w_\alpha^n P_n$, then it must vanish outside $[-1, 1]$.

In the next section we shall present a rather elementary and direct proof for Theorem 1.1. Then, in Section 3, we shall derive a short proof for the strong asymptotic result of Lubinsky and Saff for an extremal problem associated with Freud weights. With this we will provide a self contained and short proof for the most important result of the monograph [28].

In Section 4.1 we shall considerably generalize Theorem 1.1 and solve the analogous approximation problem for a large family of weights. In earlier works the approximation problem was mostly considered for concrete weights such as

Freud, Jacobi or Laguerre weights. The generalization given in Theorem 4.2 is the first general result in the subject and is far stronger than the presently existing results (e.g. it allows S_w to lie on different intervals). It also solves several open conjectures. However, the new and relatively simple *method* is perhaps the most important contribution of the present paper (Lubinsky and Saff themselves generalized Theorem 1.1 in a different direction, see e.g. [28] and Section 12). We shall first restrict our attention to the important special case given in Theorem 1.1 in order to get a simple proof for the above mentioned asymptotics (and hence for the so called Freud conjecture) and in order not to complicate our method with the technical details that are needed in the proof of Theorem 4.2 (see Section 5).

In the third part of this work we shall present a modification of the method. This will allow us to consider varying weights in the stronger sense, that we shall allow even w_n to vary with n. Recently a lot of attention has been paid to such varying weights which are connected to some interesting applications to be discussed in Chapter IV.

In essence our approximation problem can be reformulated as follows: how well can we discretize logarithmic potentials, i.e. replace them by a potential of a discrete measure which are the sums of n ($n = 1, 2, \ldots$) equal point masses (see the discussion below for the relevant concepts). The usual procedure is the following: divide the support into $n + 1$ equal parts with respect to the measure and place masses $1/n$ at these division points. This approach has proven to be sufficient and useful in many problem. However, the process introduces singularities on the support which has to be avoided in finer problems. Our method in its simplest form is a modification of the previous idea. We also divide the support into n equal parts with respect to the measure, but we use the *weight points* of these parts instead of their endpoints for placing the mass points to, then we *vertically shift* this discrete measure by an amount L_n/n where $L_n \to \infty$ is appropriately chosen. This modification will result in a dramatic increase in the speed of approximation.

$$* * * * * * * * * * * * *$$

In the rest of this introduction we shall briefly outline the results from the theory of weighted potentials that we will need in the paper.

We shall use logarithmic potentials of Borel measures. If μ is a finite Borel measure with compact support, then its logarithmic potential is defined as its convolution with the logarithmic kernel:

$$U^\mu(z) = \int \log \frac{1}{|z - t|} \, d\mu(t).$$

Let Σ be a closed subset of the real line. For simplicity we shall assume that "Σ is regular with respect to the Dirichlet problem in $\mathbf{C} \setminus \mathbf{R}$", by which we mean that every point x_0 of Σ satisfies Wiener's condition: if

$$E_n := \{x \in \Sigma \, | \, 2^{-n-1} \leq |x - x_0| \leq 2^{-n}\},$$

then

(1.2)
$$\sum_{n=1}^{\infty} \frac{n}{\log 1/\mathrm{cap}(E_n)} = \infty$$

(see the following discussion for the definition of the logarithmic capacity). In particular, this is true if Σ consists of finitely many (finite or infinite) intervals. This regularity condition is not too essential in our considerations, but it simplifies some of our proofs.

A weight function w on Σ is said to be *admissible* if it satisfies the following three conditions

(1.3)
$\quad$ (i) $\quad w$ is continuous;

$\quad$ (ii) $\quad \Sigma_0 := \{x \in \Sigma \mid w(x) > 0\}$ has positive capacity;

$\quad$ (iii) $\quad$ if Σ is unbounded, then $|x|w(x) \to 0$ as $|x| \to \infty$, $x \in \Sigma$.

We are interested in approximation of continuous functions by weighted polynomials of the form $w^n P_n$. To understand the behavior of such polynomials we have to recall a few facts from [34] and [35] about the solution of an extremal problem in the presence of a weight (often called external field).

We define $Q = Q_w$ by

(1.4)
$$w(x) =: \exp(-Q(x)).$$

Then $Q : \Sigma \to (-\infty, \infty]$ is continuous everywhere where w is positive, that is where Q is finite.

Let $\mathcal{M}(\Sigma)$ be the set of all positive unit Borel measures μ with $\mathrm{supp}(\mu) \subseteq \Sigma$, and define the weighted energy integral

(1.5)
$$\begin{aligned} I_w(\mu) &:= \iint \log[|z-t|w(z)w(t)]^{-1} d\mu(z)d\mu(t) \\ &= \iint \left[\log \frac{1}{|z-t|} + Q(z) + Q(t)\right] d\mu(z)d\mu(t). \end{aligned}$$

The classical case corresponds to choosing Σ to be compact and $w \equiv 1$ on Σ: If μ is a Borel measures with compact support on $\mathbf{R}$, then its *logarithmic energy* is defined as

$$I(\mu) := \int U^\mu(z)d\mu(z) = \iint \log \frac{1}{|z-t|} d\mu(t)d\mu(z).$$

If K is a compact set, then its *logarithmic capacity* $\mathrm{cap}(K)$ is defined by the formula

(1.6)
$$\log \frac{1}{\mathrm{cap}(K)} := \inf \{I(\mu) \mid \mu \in \mathcal{M}(K)\}.$$

Now the capacity of an arbitrary Borel set B is defined as the supremum of the capacities of compact subsets of B, and a property is said to hold *quasi-everywhere* on a set A if it holds at every point of A with the exception of points of a set of capacity zero.

The equilibrium measure (see [51] or [17]) ω_K of K is the unique probability measure ω_K minimizing the energy integrals in (1.6). Its potential has the following properties:

$$(1.7) \qquad\qquad U^{\omega_K}(z) \leq \log \frac{1}{\operatorname{cap}(K)} \quad \text{for} \quad z \in \mathbf{C},$$

$$(1.8) \qquad\qquad U^{\omega_K}(z) = \log \frac{1}{\operatorname{cap}(K)} \quad \text{for quasi-every} \quad z \in K.$$

If K is regular (which means that its complement $\mathbf{C} \setminus K$ is regular with respect to the Dirichlet problem), then we have equality for every z in (1.8).

Returning to the general case of weighted energies, the next theorem was essentially proved in [34] and [35].

Theorem A *Let w be an admissible weight on the set Σ, and let*

$$(1.9) \qquad\qquad V_w := \inf\{I_w(\mu)\,|\,\mu \in \mathcal{M}(\Sigma)\}.$$

Then the following properties are true.
   (a) V_w is finite.
   (b) There exists a unique $\mu_w \in \mathcal{M}(\Sigma)$ such that

$$I_w(\mu_w) = V_w.$$

Moreover, μ_w has finite logarithmic energy.
   (c) $S_w := \operatorname{supp}(\mu_w)$ is compact, is contained in Σ_0 (c.f. property (ii) above), and has positive capacity.
   (d) The inequality

$$U^{\mu_w}(z) \geq -Q(z) + V_w - \int Q d\mu_w =: -Q(z) + F_w$$

holds on Σ.
   (e) The inequality
$$U^{\mu_w}(z) \leq -Q(z) + F_w$$
holds for all $z \in S_w$.
   (f) In particular, for every $z \in S_w$,

$$U^{\mu_w}(z) = -Q(z) + F_w.$$

The proof is an adaptation of the classical Frostman method. In fact, in [34] and [35] property (d) was proved to hold for quasi-every $z \in \Sigma$. But the regularity of Σ implies that then the set of points where

$$\int \log \frac{1}{|z-t|} d\mu_w(t) + Q(z) \geq V_w - \int Q d\mu_w =: F_w$$

holds is dense at every point of Σ in the fine topology (see [12, Chapter 10] or [17, Chapter III]), hence the inequality in question is true at every $z \in \Sigma$ by the continuity of Q (where it is finite) and the continuity of logarithmic potentials in the fine topology.

The measure μ_w is called the *equilibrium* or *extremal measure* associated with w.

Above we have used the abbreviation

$$F_w := V_w - \int Q d\mu_w$$

for this important quantity.

We cite another theorem of H. N. Mhaskar and E. B. Saff [34, Theorem 2.1], which says that the supremum norm of weighted polynomials $w^n P_n$ lives on S_w. Let us agree that whenever we write P_n, then it is understood that the degree of P_n is at most n.

Theorem B *Let w be an admissible weight on $\Sigma \subseteq \mathbf{R}$. If P_n is a polynomial of degree at most n and*

(1.10) $$|w(z)^n P_n(z)| \leq M \quad for \quad z \in S_w,$$

then for all $z \in \mathbf{C}$

(1.11) $$|P_n(z)| \leq M \exp\left(n(-U^{\mu_w}(z) + F_w)\right).$$

Furthermore, (1.10) implies

(1.12) $$|w(z)^n P_n(z)| \leq M \quad for \quad z \in \Sigma.$$

This theorem asserts that every weighted polynomial must assume its maximum modulus on S_w. Soon we shall see that S_w is the smallest set with this property.

Theorem B is an immediate consequence of the principle of domination (see the proof of Lemma 5.1 in Section 5).

Part I

Freud weights

In the first part of the paper we shall consider exponential type (also called Freud) weights. We shall illustrate our method on them. The other purpose of this part is to give a self-contained and relatively short proof for the strong asymptotic results of Lubinsky and Saff [28].

2 Short proof for the approximation problem for Freud weights

In this section we give a short and simple proof for Theorem 1.1.

Let $Q(x) = \gamma_\alpha |x|^\alpha$, so that $w_\alpha(x) = w(x) = \exp(-Q(x))$. First we simplify the problem.

I. Obviously, it is enough to consider f's that are positive in $(-1, 1)$ and less than, say, 1. Furthermore, we know that it is sufficient to approximate on, say, $[-2, 2]$, because $w^n P_n$ tends to zero outside $[-3/2, 3/2]$ (see Theorems A and B from the introduction and the formula (3.7) in Section 3).

II. It is enough to approximate by the absolute values of weighted polynomials. In fact, if $w^n |P_n|$ uniformly tends to $\sqrt{f}$, then $w^{2n} |P_n|^2$ uniformly tends to f, and here $|P_n|^2$ is already a real polynomial. This shows our claim when the degree n is even. For odd degree one can get the statement by approximating f/w with even degree polynomials and then by multiplying through by w.

III. It is enough to show the following: for every $\epsilon > 0$ and $L > 0$ there is a continuous function g_L and for every large n polynomials Q_n of degree at most n such that with $J_\epsilon := [-1 + \epsilon, 1 - \epsilon]$

$$(2.1) \qquad w^n(x)|Q_n(x)| = \exp(g_L(x) + R_L(x)), \qquad x \in J_\epsilon,$$

where the remainder term $R_L(x)$ satisfies $|R_L(x)| \leq C_\epsilon / L$ uniformly for $x \in J_\epsilon$ with some $C_\epsilon \geq 1$ independent of L, and for every $x \in [-3, 3]$

$$(2.2) \qquad w^n(x)|Q_n(x)| \leq Dn^3,$$

where $D = D_{L,\epsilon}$ is a constant independent of n.

In fact, suppose this is true, and apply it to w^λ instead of w with some $\lambda > 1$. The corresponding extremal support is $[-\theta_\lambda, \theta_\lambda]$ with $\theta_\lambda = \lambda^{-1/\alpha}$ tending to 1 together with λ, hence, by choosing $\lambda > 1$ close to 1 and then applying the statement above to a smaller ϵ if necessary, we can see that there are polynomials $Q_{[n/\lambda]}$ of degree at most $[n/\lambda]$ such that with some g_L and R_L as above

$$w^n(x)|Q_{[n/\lambda]}(x)| = \exp(g_L(x) - (n - \lambda[n/\lambda])Q(x) + R_L(x)), \qquad x \in J_\epsilon,$$

and
$$w^n(x)|Q_{[n/\lambda]}(x)| \le D_1 n^3, \qquad x \in [-2,2].$$

Since $0 \le n - \lambda[n/\lambda] \le \lambda$, and the family of function $\{g_L - sQ \mid 0 \le s \le 1\}$ (considered on $[-1+\epsilon, 1-\epsilon])$ is compact, for every large n there are polynomials $S_{n-[n/\lambda]}$ of degree at most $n - [n/\lambda]$ such that

$$|S_{n-[n/\lambda]}(x) - f(x)\exp(-g_L(x) + (n - \lambda[n/\lambda])Q(x))|$$

$$\le \exp(-g_L(x) + (n - \lambda[n/\lambda])Q(x))/L, \qquad x \in J_{2\epsilon},$$

$$|S_{n-[n/\lambda]}(x)| \le f(x)\exp(-g_L(x) + (n - \lambda[n/\lambda])Q(x)), \qquad x \in J_{2\epsilon} \setminus J_\epsilon,$$

and
(2.3) $$|S_{n-[n/\lambda]}(x)| \le n^{-4}, \qquad x \in [-2,2] \setminus J_\epsilon.$$

Now we set $P_n = Q_{[n/\lambda]}S_{n-[n/\lambda]}$, which has degree at most n. If $\eta > 0$ is given, then choose first $\epsilon > 0$ so that the maximum of f outside $J_{2\epsilon}$ is smaller than η, then chooose $\lambda > 1$ as above, and finally choose L large enough to have $C_\epsilon/L < \eta$. Then our estimates show that for sufficiently large n the difference $|w^n|P_n| - f|$ is at most 3η on $[-2,2]$, and this is what we need to prove.

IV. Thus, we only have to verify (2.1) and (2.2).

Let us consider the so called Ullman distribution μ_w given by its density function
(2.4) $$v(t) = \frac{\alpha}{\pi} \int_{|t|}^1 \frac{u^{\alpha-1}}{\sqrt{u^2-t^2}}\, du.$$

It is well-known (see the computation in Section 3, especially (3.6) and (3.7)) that $w(x)$ and $\exp(U^{\mu_w}(x))$ differ on $[-1,1]$ only in a multiplicative constant, and elsewehere the weight $w(x)$ is smaller than $\exp(U^{\mu_w}(x))$ times this constant. Hence it is enough to show (2.1) and (2.2) with $w = w_\alpha$ replaced by $\exp(U^{\mu_w})$. In doing so we are going to use the standard discretization technique for logarithmic potentials (c.f. [42] and [28]) with some modifications, but exactly these modifications permit good approximation.

Let v be the density of the Ullman distribution μ_w (see (2.4)), and let us divide $[-1,1]$ by the points $-1 = t_0 < t_1 < \ldots < t_n = 1$ into n intervals I_j, $j = 0, 1, \ldots, n-1$ with $\mu_w(I_j) = 1/n$. Since v is continuous and positive in $(-1,1)$, there are two constants c, C (depending on ϵ) such that if $I_j \cap J_{\epsilon^2} \ne \emptyset$, then $c/n \le |I_j| \le C/n$.

Let
$$\xi_j := \frac{1}{\mu(I_j)} \int_{I_j} t\, d\mu(t) = n \int_{I_j} t\, d\mu(t)$$

be the weight point of the restriction of μ_w to I_j, and set

$$Q_n(t) = \prod_j (t - iL/n - \xi_j).$$

We claim that this choice will satisfy (2.1) and (2.2) (with w replaced by $\exp(U^{\mu_w})$).

First of all let us consider the partial derivative of $U^{\mu_w}(z)$ at $z = x + iy$ with respect to y:

$$(2.5) \qquad \frac{\partial U^{\mu_w}(z)}{\partial y} = -\int_{-1}^{1} \frac{y}{(x-t)^2 + y^2} v(t)\,dt \to \pi v(x)$$

as $y \to 0 - 0$ uniformly for $x \in J_\epsilon$ by the properties of the Poisson kernel. This, and the mean value theorem implies that

$$(2.6) \qquad U^{\mu_w}(x) - U^{\mu_w}(x - iL/n) = \frac{\pi L v(x)}{n} + o\left(\frac{L}{n}\right)$$

uniformly in $x \in J_\epsilon$. The same argument shows that

$$(2.7) \qquad |U^{\mu_w}(x) - U^{\mu_w}(x - iL/n)| = O(\frac{L}{n})$$

uniformly for $x \in \mathbf{R}$.

Actually, (2.5) and (2.6) uniformly hold on $\mathbf{R}$ because v is continuous (even at ± 1) and vanishes outside $[-1, 1]$. We shall use this fact in Section 3, but for the present purposes we keep the above formulation because in Section 4.1 we shall consider weights the density of which is not necessarily continuous around the endpoints, and it will be easier to point out the necessary changes if we work with (2.6) and (2.7).

Let $\mu_n(t) = \mu_w(t - iL/n)$, i.e. we are defining μ_n on the interval $[-1, 1] + iL/n$, which is obtained by shifting $[-1, 1]$ upwards on the plane by the amount L/n. Then the preceding two estimates tell us how far apart the two potentials U^{μ_w} and U^{μ_n} can be on $[-1, 1]$ and on $\mathbf{R}$. Next we estimate for $x \in J_\epsilon$, $x \in I_{j_0}$

$$(2.8) \qquad |\log|Q_n(x)| + nU^{\mu_n}(x)|$$

$$= \left| \sum_{j=0}^{n-1} n \int_{I_j} (\log|x - iL/n - t| - \log|x - iL/n - \xi_j|)\,d\mu_w(t) \right|.$$

Here the integrand is

$$\log\left|1 + \frac{\xi_j - t}{x - iL/n - \xi_j}\right| = \Re \log\left(1 + \frac{\xi_j - t}{x - iL/n - \xi_j}\right).$$

Since the absolute value of

$$\frac{\xi_j - t}{x - iL/n - \xi_j}, \qquad t \in I_j$$

is at most $1/2$ for large L (check this separately for $|\xi_j - t| \leq C/n$ and for the opposite case which can only occur if $I_j \cap J_{\epsilon^2} = 0$ and hence $|x - \xi_j| > \epsilon/2$ while $|I_j| < \epsilon^2$), it easily follows that then the last expression can be written in the form

$$= (\xi_j - t)\Re \frac{1}{x - iL/n - \xi_j} + O\left(\frac{|\xi_j - t|^2}{|x - iL/n - \xi_j|^2}\right),$$

and since the integral of the first term on I_j against $d\mu_w(t)$ is zero because of the choice of ξ_j, we have to deal only with the second term. For it we have the upper estimate

$$O\left(\frac{(C/n)^2}{(L/n)^2 + (c(j-j_0)/n)^2}\right)$$

if $I_j \cap J_{\epsilon^2} \neq \emptyset$ and

$$O\left(\frac{|I_j|^2}{\epsilon^2}\right)$$

otherwise (recall that $x \in J_\epsilon$), hence we can continue (2.8) as

$$\leq C_1 \sum_{k=0}^{\infty} \frac{C^2}{L^2 + c^2 k^2} + C_1 \max_j |I_j| \sum_{I_j \cap J_{\epsilon^2} = \emptyset} |I_j| \epsilon^{-2} \leq \frac{C_\epsilon}{L}$$

if n is sufficiently large.

Now

$$\log|Q_n(x)| + nU^{\mu_w}(x) = (\log|Q_n(x)| + nU^{\mu_n}(x)) + (nU^{\mu_w}(x) - nU^{\mu_n}(x)),$$

and here, by the preceding estimate, the first term is at most C_ϵ/L in absolute value, while by (2.6) the second term is $\pi v(x)L + o(L)$ uniformly in $x \in J_\epsilon$ as $n \to \infty$. This gives (2.1) (recall that we are working with $\exp(U^{\mu_w})$ instead of w).

The proof of (2.2) is standard: using the monotonicity of the logarithmic function we have for example for $x \in I_{j_0}$, $j_0 < j < n-1$ the inequality

$$\log|x - iL/n - \xi_j| \leq n \int_{I_{j+1}} \log|x - iL/n - t| d\mu_w(t),$$

and adding these and the analogous inequalities for $j < j_0$ together one can easily deduce the estimate

(2.9) $$\log|Q_n(x)| + nU^{\mu_n}(x) \leq 3\log 6 +$$

$$+ \sum_{j=j_0-1}^{j_0+1} n \int \log \frac{1}{|x - iL/n - t|} d\mu_w(t) \leq 3\log 6n/L$$

for every $x \in [-3, 3]$. This and (2.7) prove (2.2).

■

3   Strong asymptotics

The theorems of this section are not new, they can be found in the monograph [28] by D. S. Lubinsky and E. B. Saff. We closely follow many steps from [28], but we substitute the approximation part of the proof with the simple method of Section 2 which allows us to make shortcuts and simplifications, thereby significantly reducing the length of the original proof (which is scattered through about 100 pages).

First we shall consider the L^2 extremal problem and then the L^p one at the end of the section.

In order to have a complete proof we add a few standard calculations that may help the reader.

Let $w(x) = w_\alpha(x) = e^{-\gamma_\alpha |x|^\alpha}$, $\alpha > 1$ be a Freud weight on **R** normalized so that $S_w = [-1, 1]$ (this normalization is made for convenience, any other positive constant can replace γ_α on the right; for the explicit form of the constant γ_α see (3.5) below), and consider the orthonormal polynomials with respect to w^2:

$$p_n(w; x) = \gamma_n(w)x^n + \cdots$$

defined by the orthogonality relation

$$\int p_n(w; x)p_m(w; x)w^2(x)\, dx = \delta_{n,m}.$$

When $\alpha = 2$ these are the classical Hermite polynomials, for other α's G. Freud started to investigate their properties.

Let Π_n denote the set of polynomials of degree n and leading coefficients 1, i.e.

$$\Pi_n = \{x^n + \cdots\}.$$

The leading coefficient $\gamma_n(w)$ of the orthonormal polynomials p_n are closely related to a weighted extremal (minimum) problem, namely

$$(3.1) \qquad \frac{1}{\gamma_n(w)^2} = \inf_{P_n \in \Pi_n} \int P_n^2 w^2,$$

and it is one of the most important quantities related to p_n. In fact, their behavior determines the behavior of the p_n's which can also be seen by the fact, that in the recurrence formula

$$xp_n(w; x) = A_{n+1}p_{n+1}(w; x) + A_n p_{n-1}(w; x)$$

the recurrence coefficients are given by

$$A_n = \gamma_{n-1}(w)/\gamma_n(w).$$

In [6] G. Freud made two conjectures: one on the asymptotics of the largest zeros of the p_n's and another one on the recurrence coefficients. E. A Rahmanov [42] solved the first conjecture, but the second one, which claimed that

$$(3.2) \qquad \lim_{n \to \infty} n^{-1/\alpha} A_n = \frac{1}{2},$$

was open for some while, until it was settled in a series of papers [16], [29] and [27] by D. S. Lubinsky, A. Knopfmacher, P. Nevai, S. N. Mhaskar and E. B. Saff

(Freud himself verified the conjecture for $\alpha = 2, 4, 6$, and for even integers it was settled by A. Magnus [32]). This was a typical conjecture that was obviously bound to be true (already Freud new that the terms on the left are in between two positive constants, and if the limit exists then it has to be $1/2$; and there was no reason why the limit should not exist), but its proof required genuinely new tools. Shortly after settling Freud's conjecture, D. S. Lubinsky and E. B. Saff [28] proved the following strong (as opposed to (3.2), which is called ratio) asymptotics for the $\gamma_n(w)$'s themselves, which is probably one of the all-time best results in the theory of orthogonal polynomials:

$$(3.3) \qquad \lim_{n \to \infty} \gamma_n(w) \pi^{1/2} 2^{-n} e^{-n/\alpha} n^{(n+1/2)/\alpha} = 1.$$

Below we shall present a relatively short proof for (3.3) that utilizes the approximation technique in Section 2. The original proof is scattered through the monograph [28] and is quite long.

Let us start with the Ullman distribution given by its density

$$(3.4) \qquad v(t) = \frac{\alpha}{\pi} \int_{|t|}^{1} \frac{u^{\alpha-1}}{\sqrt{u^2 - t^2}} \, du$$

on $[-1, 1]$. For its potential we have (writing instead of the measure its density as a parameter in U) by switching the order of integration

$$-U^v(x) = \int_0^1 \alpha u^{\alpha-1} \frac{1}{\pi} \int_{-u}^{u} \frac{\log|x - t|}{\sqrt{u^2 - t^2}} \, dt \, du.$$

The expression after $u^{\alpha-1}$ is nothing else than the negative of the equilibrium potential of the interval $[-u, u]$, hence it equals $\log u - \log 2$ if $|x| \leq u$ and $\log|x + \sqrt{x^2 - u^2}| - \log 2$ if $|x| > u$. Thus, for $-1 \leq x \leq 1$ we can continue the above equality as

$$= \; -\log 2 + \int_{|x|}^{1} \alpha u^{\alpha-1} \log u \, du + \int_0^{|x|} \alpha u^{\alpha-1} \log|x + \sqrt{x^2 - u^2}| \, du$$

$$= \; -\log 2 - \frac{1}{\alpha} + |x|^\alpha \left(\frac{1}{\alpha} + \int_0^1 \alpha v^{\alpha-1} \log(1 + \sqrt{1 - v^2}) \, dv \right).$$

Integration by parts yields that the last integral is

$$= \int_0^1 \frac{v^{\alpha-1}}{1 + \sqrt{1 - v^2}} \frac{v^2}{\sqrt{1 - v^2}} dv = \int_0^1 \frac{v^{\alpha-1}}{\sqrt{1 - v^2}} dv - \frac{1}{\alpha},$$

where we used the identity $v^2 = (1 - \sqrt{1 - v^2})(1 + \sqrt{1 - v^2})$. Since

$$(3.5) \qquad \gamma_\alpha := \int_0^1 \frac{v^{\alpha-1}}{\sqrt{1 - v^2}} \, dv = \Gamma(\frac{\alpha}{2})\Gamma(\frac{1}{2})/(2\Gamma(\frac{\alpha}{2} + \frac{1}{2})),$$

we finally get for $x \in [-1, 1]$

$$(3.6) \qquad U^v(x) = -\gamma_\alpha |x|^\alpha + \log 2 + \frac{1}{\alpha}.$$

Let now $|x| > 1$, $x \in \mathbf{R}$. By symmetry we can assume $x > 1$. Exactly as above

$$U^v(x) = \log 2 - \int_0^1 \alpha u^{\alpha-1} \log(x + \sqrt{x^2 - u^2})du,$$

and by differentiation we get

$$(U^v(x))' = -\int_0^1 \frac{\alpha u^{\alpha-1}}{\sqrt{x^2 - u^2}}du = -\alpha\gamma_\alpha |x|^{\alpha-1} + \int_{1/x}^1 \frac{\alpha u^{\alpha-1}}{\sqrt{1 - u^2}}\, du.$$

This tells us first of all that outside $[-1,1]$ we have on $\mathbf{R}$

$$(3.7) \qquad U^v(x) > -\gamma_\alpha |x|^\alpha + \log 2 + \frac{1}{\alpha},$$

and so by (3.6) and Lemma 5.1 to be proven in Section 5 we can conclude that $d\mu_w(t) = v(t)dt$ and
$$(3.8) \qquad F_w = \log 2 + 1/\alpha.$$

It also follows that for $|x| \in (1, 2)$

$$(U^v(x))' + \alpha\gamma_\alpha x^{\alpha-1} \sim ||x| - 1|^{1/2},$$

where $\sim$ indicates that the ratio of the two sides lies in between two absolute constants (in the range of the arguments indicated), and so

$$(3.9) \qquad (U^{\mu_w}(x) - F_w) + \gamma_\alpha |x|^\alpha \sim ||x| - 1|^{3/2}$$

for $|x| \in (1, 2)$, and
$$(3.10) \qquad (U^{\mu_w}(x) - F_w) + \gamma_\alpha |x|^\alpha \sim |x|^\alpha$$

when $|x| > 2$.

This and Theorem B of the introduction easily imply the following inequality of D. S. Lubinsky [25]: for $\rho_n = 1 + n^{-7/12}$

$$(3.11) \qquad \sup_{\deg P_n \leq n} \left(\int_{-\infty}^\infty e^{-\gamma_\alpha 2n|x|^\alpha} P_n^2(x)\, dx \Big/ \int_{-\rho_n^{1/\alpha}}^{\rho_n^{1/\alpha}} e^{-\gamma_\alpha 2n|x|^\alpha} P_n^2(x)\, dx \right)$$

$$= 1 + o(1)$$

as $n \to \infty$ (for completeness we shall give a short proof for (3.11) at the end of this section). The awkward looking $\rho_n^{1/\alpha}$ in the limits of integration on the left is just to match the proof below, we shall only need that it is larger than $1 + cn^{-7/12}$ with some $c > 0$.

After these preliminaries let us return to (3.3). Let $\varphi(x) = 1 - x^2$. We need the following formula of S. N. Bernstein (see [2, pp. 250–254] or [28, p. 111]):

Let R_{2q} be a polynomial of degree $2q$, positive on $(-1,1)$ with possibly simple zeros at ± 1. Then for $n \geq q$

(3.12) $$\left(\inf_{P_n \in \Pi_n} \int_{-1}^{1} \frac{\varphi^{1/2}}{R_{2q}} P_n^2 \right)^{1/2} = \pi^{1/2} 2^{-n} \exp\left(\frac{1}{\pi} \int_{-1}^{1} \frac{\log(\varphi^{1/4}/R_{2q}^{1/2})}{\varphi^{1/2}} \right).$$

In what follows let us abbreviate the geometric mean appearing on the right as $G[\varphi^{1/4}/R_{2q}^{1/2}]$, i.e.

(3.13) $$G[V] = \exp\left(\frac{1}{\pi} \int_{-1}^{1} \frac{\log V(x)}{\sqrt{1-x^2}} \, dx \right).$$

3.1 The upper estimate

Let $\rho_n = 1 - n^{-2/3}$, and let us carry out the substitution $x = \rho_n^{1/\alpha} n^{1/\alpha} y$ in the integrals in (3.1), and then restrict the integrals to $[-1,1]$. We get

(3.14) $$\frac{1}{\gamma_n(w)^2} \geq n^{(2n+1)/\alpha} \rho_n^{(2n+1)/\alpha} \inf_{P_n \in \Pi_n} \int_{-1}^{1} e^{-\gamma_\alpha \rho_n 2n |x|^\alpha} P_n^2(x) \, dx.$$

We are going to show with the method of Section 2 that there are polynomials H_n of degree at most n such that if

(3.15) $$h_n(x) = e^{-\gamma_\alpha \rho_n n |x|^\alpha} |H_n(x)| (1-x^2)^{-1/4},$$

then
(3.16) $$h_n(x) \geq 1 \qquad \text{for} \quad x \in [-1,1]$$

and
(3.17) $$\lim_{n \to \infty} G[h_n] = 1.$$

Then we will have by (3.14) and (3.16)

$$\frac{1}{\gamma_n(w)^2} \geq n^{(2n+1)/\alpha} \rho_n^{(2n+1)/\alpha} \inf_{P_n \in \Pi_n} \int_{-1}^{1} \frac{\varphi^{1/2}}{|H_n|^2} P_n^2,$$

and so by Bernstein's formula

(3.18) $$\frac{1}{\gamma_n(w)^2} \geq n^{(2n+1)/\alpha} \rho_n^{(2n+1)/\alpha} \pi 2^{-2n} \left(G[\varphi^{1/4}/|H_n|] \right)^2.$$

But here
$$\varphi^{1/4}(x)/|H_n(x)| = e^{-\gamma_\alpha \rho_n n |x|^\alpha}/h_n(x)$$

and
(3.19) $$\gamma_\alpha \frac{1}{\pi} \int_{-1}^{1} \frac{|x|^\alpha}{\sqrt{1-x^2}} \, dx = \frac{1}{\alpha},$$

which, together with (3.17) imply that $\rho_n^{2n/\alpha}$ times the geometric mean on the right hand side of (3.18) has the form

$$(1+o(1))e^{-2n/\alpha} \exp\left((2n/\alpha)((1-\rho_n) + \log \rho_n) \right) = (1+o(1))e^{-2n/\alpha},$$

and this proves that

$$\limsup_{n \to \infty} \gamma_n(w) \pi^{1/2} 2^{-n} e^{-n/\alpha} n^{(n+1/2)/\alpha} \le 1.$$

Thus, everything boils down to the existence of polynomials H_n with properties (3.16) and (3.17).

We follow the proof in Section 2, but now we need somewhat finer analysis around the endpoints. By symmetry, we can restrict our attention to the left endpoint -1. For the Ullman distribution (3.4) it immediately follows that

$$v(t) \sim (1 - t^2)^{1/2} \qquad \text{as} \quad t \to \pm 1,$$

and this property alone implies for the I_k's and ξ_k's of Section 2 that for $0 \le k \le n/2$

$$(3.20) \qquad 1 + \xi_k \sim \left(\frac{k+1}{n} \right)^{2/3}, \qquad |I_k| \sim \frac{1}{(k+1)^{1/3} n^{2/3}},$$

and analogous estimates hold for $k \ge n/2$. Hence for $j \ne j_0$

$$\text{dist}(\xi_j, I_{j_0}) \sim \frac{1}{n^{2/3}} \sum_{k \in [j_0, j]} \frac{1}{(k+1)^{1/3}} \sim \frac{|j^{2/3} - j_0^{2/3}|}{n^{2/3}}.$$

Since for $x \in I_{j_0}$, the absolute value of the the j-th, $j \ne j_0$, term in the sum in (2.8) is at most

$$\left(\frac{|I_j|}{\text{dist}(\xi_j, I_{j_0})} \right)^2,$$

(see the argument after (2.8)), it follows from the preceding estimate that

$$(3.21) \qquad \sum_{j \ne j_0} n \left| \int_{I_j} (\log |x - iL/n - t| - \log |x - iL/n - \xi_j|) \, d\mu_w(t) \right|$$

$$\le C \sum_{j \ne j_0} \left(\frac{1}{(j+1)^{1/3} n^{2/3}} \right)^2 \Big/ \left(\frac{|j^{2/3} - j_0^{2/3}|}{n^{2/3}} \right)^2 = O(1).$$

Thus, it has left to estimate the j_0-th term in (2.8). Its absolute value is obviously bounded by

$$\frac{1}{2} \log \frac{(L/n)^2 + |I_{j_0}|^2}{(L/n)^2} \le \frac{1}{2} \log \left(1 + \frac{1}{L^2} O\left(\left(\frac{n}{j_0 + 1} \right)^{2/3} \right) \right) \le \frac{1}{2} \log \frac{1}{1 - x^2}$$

if L is sufficiently large (recall that $x \in I_{j_0}$, hence

$$\left(\frac{n}{j_0 + 1} \right)^{2/3} \le C \frac{1}{1 - x^2}$$

by (3.20)).

Taking into account (2.6), which, as we have seen in Section 2, uniformly holds on $\mathbf{R}$, we get that the polynomials Q_n from Section 2 satisfy

$$(3.22) \qquad \frac{1}{C}\sqrt{1-x^2} \le e^{-\gamma_\alpha n|x|^\alpha}|Q_n(x)|e^{nF_w} \le \frac{C}{\sqrt{1-x^2}}$$

uniformly in $x \in [-1, 1]$ and n with some constant C (recall (3.8) and (3.6)), and

$$(3.23) \qquad \lim_{n\to\infty} e^{-\gamma_\alpha n|x|^\alpha}|Q_n(x)|e^{nF_w} = e^{L\pi v(x)}$$

uniformly on compact subsets of $(-1, 1)$.

Next we improve the estimate (3.22) for x lying close to ± 1. The proof of (3.21) and (3.22) easily yields that there is a $c > 0$ such that

$$(3.24) \quad e^{-\gamma_\alpha n|x|^\alpha}|Q_n(x)|e^{nF_w} \ge c \qquad \text{for all } n \text{ and } 0 < 1 - |x| \le cn^{-2/3}.$$

In fact, recall that, say, $|I_0| \sim (1 + \xi_0) \sim n^{-2/3}$, and so (c.f. (3.21) and the argument after (2.8) in Section 2)

$$n\left|\int_{I_0} (\log|x - iL/n - t| - \log|x - iL/n - \xi_0|)\,d\mu_w(t)\right|$$

$$= n\int_{I_0}\left((\xi_0 - t)\Re\frac{1}{x - iL/n - \xi_0} + O\left(\frac{|\xi_0 - t|^2}{|x - iL/n - \xi_0|^2}\right)\right)d\mu_w(t)$$

$$= n\int_{I_0} O(1)d\mu_w(t) \le C'$$

provided $0 < 1 - |x| \le cn^{-2/3}$ and c is sufficiently small.

For later purposes let us record here that the proof gives also the following:

$$(3.25) \qquad \frac{1}{C} \le \exp(nU^{\mu_w}(x))|Q_n(x)| \le C$$

uniformly on $\mathbf{R} \setminus [-1, 1]$.

Now let us remove the two zeros from Q_m which have the smallest and the largest real parts, respectively. On the interval $[-1 + cn^{-2/3}, 1 - cn^{-2/3}]$ this introduces a factor $\sim c_1/(1 - x^2)^{-1}$, hence for the so modified polynomial Q_n^* we get from (3.22), (3.23) and (3.24) that

$$(3.26) \qquad \frac{1}{C} \le e^{-\gamma_\alpha n|x|^\alpha}|Q_n^*(x)|e^{nF_w} \le \frac{C}{(1-x^2)^{3/2}}$$

holds with some constant C uniformly in $x \in [-1, 1]$ and n, and

$$(3.27) \qquad \lim_{n\to\infty} e^{-\gamma_\alpha n|x|^\alpha}|Q_n^*(x)|e^{nF_w} = \frac{e^{L\pi v(x)}}{1-x^2}$$

uniformly on compact subsets of $(-1, 1)$.

After these preparatory steps let us return to (3.16) and (3.17). We apply the preceding estimates for $Q^*_{[n\rho_n]}$, which has degree at most $[n - n^{1/3}]$ to conclude

(3.28) $$\frac{1}{C} \le e^{-\gamma_\alpha \rho_n n |x|^\alpha} |Q^*_{[n\rho_n]}(x)| e^{n\rho_n F_w} \le \frac{C}{(1 - x^2)^{3/2}}$$

and

(3.29) $$\lim_{n \to \infty} e^{-\gamma_\alpha [\rho_n n]|x|^\alpha} |Q^*_{[n\rho_n]}(x)| e^{[n\rho_n]F_w} = \frac{e^{L\pi v(x)}}{1 - x^2}$$

uniformly on compact subsets of $(-1, 1)$. Then it is easy to find polynomials $R_{n-[n\rho_n]}$ of degree at most $n - [n\rho_n] \ge n^{1/3}$ such that with $H_n = Q_{[n\rho_n]}R_{n-[n\rho_n]}$ both properties (3.16) and (3.17) are satisfied (use exactly as in Section 2 the fact that the family of functions $\{-sQ \mid 0 \le s \le 1\}$ considered on $[-1, 1]$ is compact).

The point is that in the definition of h_n in (3.15) the factor $(1 - x^2)^{-1/4}$ appears, which only improves the lower estimate in (3.28), and so (3.16) is easy to achieve. To achieve (3.17) at the same time is a simple approximation procedure if we use the upper estimate in (3.28) and the asymptotic relation (3.29).   ∎

3.2   The lower estimate

The proof of the lower estimate is very similar to the above argument. In fact, let now $\rho_n = 1 + n^{-7/12}$, and let us carry out the substitution $x = \rho_n^{1/\alpha} n^{1/\alpha} y$ in the integrals in (3.1). The result is

$$\frac{1}{\gamma_n(w)^2} = n^{(2n+1)/\alpha} \rho_n^{(2n+1)/\alpha} \inf_{P_n \in \Pi_n} \int_{-\infty}^{\infty} e^{-\gamma_\alpha \rho_n 2n |x|^\alpha} P_n^2(x)\, dx.$$

Since now we want to prove a lower estimate, we cannot restrict the integral to $[-1, 1]$, rather we need the so called infinite-finite range inequality (3.11) which tells us that the part of the integrals of weighted polynomals away form the extremal support is negligible. In the above integral not the weight w^{2n} but $(w^{\rho_n})^{2n}$ appears, and the corresponding extremal measure has support $S_{w^{\rho_n}} = [-\rho_n^{-1/\alpha}, \rho_n^{-1/\alpha}]$. Thus, if we apply (3.11) then we can conclude that, by restricting the integrals to $[-1, 1]$ we introduce only a constant that tends to zero, i.e.

(3.30) $$\frac{1}{\gamma_n(w)^2} =$$

$$(1 + o(1)) n^{(2n+1)/\alpha} \rho_n^{(2n+1)/\alpha} \inf_{P_n \in \Pi_n} \int_{-1}^{1} e^{-\gamma_\alpha \rho_n 2n |x|^\alpha} P_n^2(x)\, dx.$$

Actually, the infinite-finite range inequality (3.11) has to be applied to the intervals $[-\rho_n^{-1/\alpha}, \rho_n^{-1/\alpha}]$ and $[-1, 1]$ rather than to $[-1, 1]$ and $[-\rho_n^{1/\alpha}, \rho_n^{1/\alpha}]$, which only means a linear transformation not introducing any new constant in the ratios in question. In general, this linear transformation $x \rightarrow \rho_n^{1/\alpha} y$ introduces in our formulae only a constant that tends to 1 as $n \rightarrow \infty$, hence in what follows we shall use it without explicit mentioning.

Let us now consider the weight $w^{\rho_n^2}$ for which $S_{w^{\rho_n^2}} = [-\rho_n^{-2/\alpha}, \rho_n^{-2/\alpha}]$. By (3.22), (3.23) and (3.25) there are polynomials $Q_{[n/\rho_n]}$ of degree at most $[n/\rho_n]$ such that with $w_n = w^{\rho_n^2}$

$$(3.31)\quad \frac{1}{C}(\rho_n^{-4/\alpha} - x^2)^{1/2} \leq e^{-\gamma_\alpha \rho_n n |x|^\alpha}|Q_{[n/\rho_n]}(x)|e^{n/\rho_n F_{w_n}} \leq \frac{C}{(\rho_n^{-4/\alpha} - x^2)^{1/2}}$$

uniformly in $x \in [-\rho_n^{-2/\alpha}, \rho_n^{-2/\alpha}]$ and n,

$$(3.32)\qquad\qquad \frac{1}{C} \leq \exp\left(\frac{n}{\rho_n}U^{\mu_{w_n}}(x)\right)|Q_{[n/\rho_n]}(x)| \leq C$$

uniformly in n and $x \notin [-\rho_n^{-2/\alpha}(1 - cn^{-2/3}), \rho_n^{-2/\alpha}(1 - cn^{-2/3})]$, and

$$(3.33)\qquad \lim_{n\to\infty} e^{-\gamma_\alpha \rho_n^2 [n/\rho_n]|x|^\alpha}|Q_{[n/\rho_n]}(x)|e^{[n/\rho_n]F_{w_n}} = e^{L\pi v(x)}$$

uniformly on compact subsets of $(-1, 1)$. On applying (3.9) we can conclude from (3.32) that

$$\int_{\rho_n^{-2/\alpha} \leq |x| \leq 1} \frac{\log\left(e^{-\gamma_\alpha \rho_n n |x|^\alpha}|Q_{[n/\rho_n]}(x)|e^{(n/\rho_n)F_{w_n}}(1 - x^2)^{1/4}\right)}{\sqrt{1 - x^2}}\, dx$$

$$= O\left(n(\rho_n - 1)^{3/2}(\rho_n - 1)^{1/2} + (\rho_n - 1)^{1/2}\log\frac{1}{\rho_n - 1}\right)$$

$$= O(n^{-1/6}) = o(1)$$

by the choice of the ρ_n's, which estimate is used in (3.36) below.

From (3.31) and (3.33) it easily follows that we can multiply this $Q_{[n/\rho_n]}$ by a suitable $R_{n-1-[n/\rho_n]}$ of degree at most $n - 1 - [n/\rho_n] \geq \frac{1}{2}n^{5/12}$ to get a H_{n-1} with the following properties: H_{n-1} does not vanish on $(-1, 1)$, if

$$(3.34)\qquad h_n(x) = e^{-\gamma_\alpha \rho_n n |x|^\alpha}|H_{n-1}(x)(1 - x^2)^{1/2}|(1 - x^2)^{-1/4},$$

then
$$(3.35)\qquad\qquad h_n(x) \leq 1 \qquad \text{for}\quad x \in [-1, 1],$$

and
$$(3.36)\qquad\qquad \lim_{n\to\infty} G[h_n] = 1.$$

From here the proof is the same as in the case of the upper estimate: set

$$R_{2n}(x) = |H_{n-1}(x)|^2(1 - x^2)$$

into Bernstein's formula, use (3.35) and (3.36) instead of (3.16) and (3.17), and reverse the corresponding inequalities. In the end we obtain

$$\liminf_{n \to \infty} \gamma_n(w) \pi^{1/2} 2^{-n} e^{-n/\alpha} n^{(n+1/2)/\alpha} \geq 1,$$

and the proof is complete. ∎

We have promised a proof for (3.11). First we need a crude Nikolskii–type inequality. By approximating first $\gamma_\alpha n |x|^\alpha$ on $[-2, 2]$ by some polynomials T_{n^2} of degree n^2 with error $1/n$ (this is possible by Jackson's theorem), then taking the n^2-th partial sum of the Taylor expansion of e^{-x}, and then substituting here $T_n(x)$ for x, we can get a polynomial R_{n^4} of degree at most n^4 such that uniformly in n and $x \in [-2, 2]$

$$R_{n^4}(x) \sim w^n(x).$$

Using this and the classical Nikolskii inequality

$$\|Q_n\|_{[-1,1]} \leq C n^{1/2} \|Q_n\|_{L^2[-2,2]}$$

with $Q_{n^5} = R_{n^4} P_n$ we get for any P_n

$$
\begin{aligned}
\|w^n P_n\|_{[-1,1]} &\leq C\|R_{n^4} P_n\|_{[-1,1]} \leq C n^3 \|R_{n^4} P_n\|_{L^2[-2,2]} \\
&\leq C n^3 \|w^n P_n\|_{L^2[-2,2]} \leq C n^3 \|w^n P_n\|_{L^2}
\end{aligned}
$$

(actually the best exponent on the right is $1/2\alpha$, see [40], but we will not need this). This, Theorem B from the introduction and (3.9)–(3.10) easily imply with some $c > 0$

$$\int_{|x| \geq \rho_n^{1/\alpha}} e^{-\gamma_\alpha 2n |x|^\alpha} P_n^2(x)\, dx$$

$$\leq C n^3 \|w^n P_n\|_{L^2}^2 \left(\int_{\rho_n}^{2} e^{-2cn(x-1)^{3/2}}\, dx + \int_{2}^{\infty} e^{-cnx}\, dx \right) \leq \frac{1}{n} \|w^n P_n\|_{L^2}^2,$$

and this is (3.11). ∎

3.3   The L^p case

In [28] Lubinsky and Saff considered the weighted L^p-extremal problem

$$E_{n,p}(w) := \inf_{P_n \in \Pi_n} \|w P_n\|_{L^p}.$$

Note that when $p = 2$, then this is the same as the one in (3.1). They proved the following generalization of (3.3): if $w(x) = \exp(-\gamma_\alpha |x|^\alpha)$, then

$$(3.37) \qquad \lim_{n \to \infty} E_{n,p}(w)\sigma_p^{-1} 2^{n-1+1/p} e^{n/\alpha} n^{-(n+1/p)/\alpha} = 1,$$

where

$$\sigma_p = \left(\Gamma(1/2)\Gamma((p+1)/2)/\Gamma(p/2+1) \right)^{1/p}.$$

In L^p Bernstein's formula takes the following form: Let $1 \le \infty$, and R_{2q} a polynomial of degree $2q$ which is positive on $(-1,1)$ with possibly simple zeros at ± 1. Then for $n \ge q$

$$\left(\inf_{P_n \in \Pi_n} \int_{-1}^1 \frac{\varphi^{p/2-1/2}}{R_{2q}^{p/2}} |P_n|^p \right)^{1/p}$$

$$(3.38) \qquad = \sigma_p 2^{-n+1-1/p} \exp \left(\frac{1}{\pi} \int_{-1}^1 \frac{\log(\varphi^{1/2-1/2p}/R_{2q}^{1/2})}{\varphi^{1/2}} \right),$$

i.e.

$$E_{n,p}(\varphi^{1/2-1/2p}/R_{2q}^{1/2}) = \sigma_p 2^{-n+1-1/p} G\left[\varphi^{1/2-1/2p}/R_{2q}^{1/2} \right].$$

Using this formula instead of (3.12) we can copy the proof of (3.3) and we can get (3.37) with minor modifications. For example, in the proof of the upper estimate (which correponds to the lower estimate on $\gamma_n(w)$ discussed in Section 3.2) we have to use an obvious modification in the infinite-finite range inequality (3.11), and change (3.15) to

$$h_n(x) = e^{-\gamma_\alpha \rho_n n |x|^\alpha} |H_n(x)| (1 - x^2)^{(-1+1/p)/2},$$

resp.

$$h_n(x) = e^{-\gamma_\alpha \rho_n n |x|^\alpha} |H_{n-1}(x)(1 - x^2)^{1/2}| (1 - x^2)^{(-1+1/p)/2},$$

(c.f. (3.15) and (3.34)) for which (3.35) and (3.36) can be achieved exactly as before.

In a similar manner, only minor changes have to be done in the lower estimate (which correponds to the upper estimate on $\gamma_n(w)$ discussed in Section 3.1).

Part II

Approximation with general weights

In the second part of the paper we will consider the approximation problem for general weights.

4 A general approximation theorem

4.1 Statement of the main results

Let Σ be a regular closed subset of the real line and w an admissible weight on Σ. We consider the problem of approximating a continuous function f by weighted polynomials $w^n P_n$. First we show that every such function must vanish outside the support S_w of the extremal measure μ_w (see the introduction).

Theorem 4.1 *Let us suppose that there is a sequence $\{P_n\}$ of polynomials of corresponding degree $n = 1, 2 \ldots$ such that $w^n(x)P_n(x)$ uniformly converges to a function f on $S_w \cup \{x_0\}$, $x_0 \notin S_w$. Then $f(x_0) = 0$.*

Let us note that the mere boundedness of $w^n P_n$ on S_w does not necessarily imply that the sequence $\{w^n(x_0)P_n(x_0)\}$ converges to zero (for $x_0 \notin S_w$). A counterexample is furnished by the weight w which is 1 on $[-1, 1]$ and equals $(x + \sqrt{x^2 - 1})^{-1}$ on $(1, 2]$ (consider the energy problem from the introduction on $\Sigma = [-1, 2]$), and the Chebyshev polynomials

$$T_n(x) = \frac{1}{2} \left((x + \sqrt{x^2 - 1})^n + (x - \sqrt{x^2 - 1})^n \right).$$

In this case μ_w is the arcsine measure $((\pi\sqrt{1 - x^2})^{-1} dx)$ on $[-1, 1]$, and it is obvious that $w^n T_n$ is bounded on $S_w = [-1, 1]$ but $w^n(x_0)T_n(x_0) \geq 1/2$ (and $w^n(x_0)T_n(x_0) \to 1/2$) for all $x_0 \in (1, 2]$.

Next we turn to conditions guaranteeing approximation. Let $O \subset \Sigma$ be an open subset of the real line. The space of continuous real functions that vanish outside O will be denoted by $C_0(O)$.

Definition. We say that w has the *approximation property on the open set O* if for every $f \in C_0(O)$ there is a sequence of polynomials $\{P_n\}_{n=1}^\infty$ such that $w^n P_n$ converges uniformly to f on Σ.

Thus, what we have said above implies that we can hope for the approximation property on an open set O only if $O \subseteq S_w$, that is O should be part of the interior $\text{Int}(S_w)$ of S_w. Our main result is that on the other hand, if μ_w has continuous and positive density function on the interior of S_w, then w does have the approximation property on $\text{Int}(S_w)$.

To formulate the main result of the paper we introduce the following definition:

Definition. Let S^w denote the set of those points x_0 where the equilibrium measure μ_w has continuous and positive density, that is

$$d\mu_w(t) = v_w(t)\, dt$$

in a neighborhood of x_0, and the density function v_w is continuous and positive in a neighborhood of x_0. This S^w is called the *restricted support* of μ_w.

Thus, if μ has positive and continuous density on $\mathrm{Int}(\mathcal{S}_w)$, then $S^w = \mathrm{Int}(\mathcal{S}_w)$. On the other hand, if at x_0 we have $v_w(x_0) = 0$, then this x_0 *does not* belong to the restricted support.

Theorem 4.2 *Let w be an admissible weight on $\Sigma \subseteq \mathbf{R}$. Then w has the approximation property on the restricted support S^w. In particular, if μ_w has continuous and positive density on $\mathrm{Int}(\mathcal{S}_w)$, then every continuous function that vanishes outside $\mathrm{Int}(\mathcal{S}_w)$ can be uniformly approximated on Σ by weighted polynomials of the form $w^n P_n$, where the degree of P_n is at most n.*

As a corollary of Theorem 4.2 we get

Theorem 4.3 *Suppose that $\Sigma \subseteq \mathbf{R}$ consists of finitely many disjoint intervals I_j, and w is an admissible weight of class $C^{1+\epsilon}$ for some $\epsilon > 0$ such that $Q = \log 1/w$ is convex on every I_j. Then w has the approximation property on the interior of the support $\mathcal{S}_w$.*

From the proofs of Theorems 4.2 and 4.3 the following result immediately follows.

Theorem 4.4 *Suppose that w is an admissible weight of class $C^{1+\epsilon}$ for some $\epsilon > 0$. Then w has the approximation property on the union of the interiors of the supports $\mathcal{S}_{w^\lambda}$, $\lambda > 1$.*

Now we show that Theorem 4.2 is sharp in a certain sense. To illustrate Theorem 4.2 let us consider the case when $\mathcal{S}_w = [-1, 1]$, and μ_w has continuous density v_w in $(-1, 1)$. We have seen that if this density is positive in $(-1, 1)$, then on $\mathrm{Int}(\mathcal{S}_w) = (-1, 1)$ the weight w has the approximation property, and in general this is the largest set where approximation is possible. Now what happens if v_w vanishes at a single point, say at $x = 0$? We will construct an example in Section 7.1 where this single zero prohibits approximation in a very strong sense, namely approximation is possible only for functions that vanish at the origin.

Example 4.5 *There exists a weight w such that the support of the corresponding extremal measure is $[-1, 1]$, this measure has continuous density in $(-1, 1)$ which is positive everywhere except at 0, and still no function that is nonzero at 0 can be approximated by weighted polynomials of the form $w^n P_n$.*

Hence, in this case the largest set for the approximation problem of the present section is the restricted support $(-1,0) \cup (0,1)$. This shows that, in general, on no larger set than the restricted support can w have the approximation property; in other words, Theorem 4.2 cannot be improved.

Theorem 4.2 does not tell us if approximation is possible on $\text{Int}(\mathcal{S}_w)$ if the density v_w vanishes there. Example 4.5 shows that such internal zeros may prevent approximation, but this does not rule out the possibility of the approximation property on the whole $\text{Int}(\mathcal{S}_w)$ for a concrete function. Our next example together with Example 4.5 shows that, indeed, the situation is very delicate, approximation in the presence of internal zeros depends on the weight in a subtle way.

Example 4.6 *There exists a weight w on $[-1,1]$ such that the support of the corresponding extremal measure μ_w is $[-1,1]$, μ_w has continuous density in $(-1,1)$ which vanishes at the origin, and still every continuous f that is zero at ± 1 can be uniformly approximated by weighted polynomials of the form $w^n P_n$.*

4.2 Examples and historical notes

The type of approximation we are discussing has evolved from G. G. Lorentz' incomplete polynomials . Lorentz [23] studied polynomials on $[0,1]$ that vanish at zero with high order. That is, he considered polynomials of the form

$$(4.1) \qquad P_n(x) = \sum_{k=s_n}^{n} a_k x^k,$$

and he verified that if $s_n/n \to \theta$ and the P_n's are bounded on $[0,1]$, then $P_n(x)$ tends to zero uniformly on compact subsets of $[0, \theta^2)$, and it was shown in [46] that $[0, \theta^2)$ is the largest set with this property. Although here there is no fixed weight, the resemblance to weighted polynomials $w^n P_n$ with $w(x) = x^{\theta/(1-\theta)}$ is obvious, and in fact, it is easy to transform results concerning incomplete polynomials into analogous ones concerning such weighted polynomials, and vice versa.

In our terminology this result means that the support of the extremal measure for the weight $w(x) = x^{\theta/(1-\theta)}$, $\Sigma = [-1,1]$ is $[\theta^2, 1]$. The corresponding approximation problem, namely that every $f \in C[0,1]$ that vanishes on $[0, \theta^2)$ is the uniform limit of polynomials of the form (4.1), was independently proved by v. Golitschek [8] and Saff and Varga [46].

In [44] E. B. Saff generalized the problem to exponential weights of the form $\exp(-c|x|^\alpha)$, $\alpha > 1$. With H. N. Mhaskar they proved in [36] that in this case the extremal support is

$$(4.2) \qquad S_w = [-\gamma_\alpha^{1/\alpha} c^{-1/\alpha}, \gamma_\alpha^{1/\alpha} c^{-1/\alpha}]$$

(c.f. (1.1)), and they also determined the extremal measure (given by the Ullman distribution (2.4)) in this case. In [44] Saff conjectured that every continuous function that vanishes outside (4.2) can be uniformly approximated by weighted polynomials $w^n P_n$. This was shown to be true in the special case $\alpha = 2$ in [37] by Mhaskar and Saff, and, as we have mentioned in Section 1, by Lubinsky and Saff [29] for all $\alpha > 1$. The missing range $0 < \alpha \leq 1$ was settled by D. S. Lubinsky and the author [30] proving that approximation is still possible if $\alpha = 1$, and for $\alpha < 1$ a necessary and sufficient condition that f be the uniform limit of weighted polynomials $w_\alpha^n P_n$ is that f vanishes outside $[-1, 1]$ *and at the origin*. It was also proven there (for the case $\alpha > 1$) that even if we consider f only on the interval $[-1, 1]$, it must vanish at the endpoints in order to be the uniform limit of weighted polynomials, i.e. approximation is not possible up to the endpoint for nonvanishing functions. More precisely, the following exact range for the approximation was established: Suppose that for $n \geq 1$ we are given closed intervals J_n symmetric about 0, and polynomials P_n of degree $\leq n$ such that

$$\lim_{n \to \infty} \|w_\alpha^n P_n - 1\|_{J_n} = 0.$$

Then there exists a sequence $\{\rho_n\}_{n=1}^\infty$ with

(4.3) $$\lim_{n \to \infty} \rho_n = \infty,$$

such that for infinitely many n

$$J_n \subset [-1 + \rho_n n^{-2/3}, 1 - \rho_n n^{-2/3}].$$

Conversely, if $\{\rho_n\}_{n=1}^\infty$ is a sequence satisfying (4.3), then for every continuous $f \in C[-1, 1]$ there exist polynomials P_n of degree at most n such that

$$\lim_{n \to \infty} \|w_\alpha^n P_n - f\|_{[-1+\rho_n n^{-2/3}, 1 - \rho_n n^{-2/3}]} = 0,$$

and

$$\sup_n \|w_\alpha^n P_n\|_{\mathbf{R}} < \infty.$$

In [37] the conjecture was made that even for general continuous weights w approximation by weighted polynomials $w^n P_n$ is possible for an f if and only if f vanishes outside S_w. The necessity of the condition follows from Theorem 4.1. Its sufficiency is not true, a counterexample is furnished by Example 4.5 or by $w(x) = \exp(-|x|^\alpha)$, $\alpha < 1$. In [3] the weaker conjecture was stated that at least for the case when $Q = \log 1/w$ is convex, a necessary and sufficient condition for approximation is the same as before, that is that the function vanishes outside S_w. This conjecture of Borwein and Saff follows from Theorem 4.3 under the minimal smoothness assumption that Q is a $C^{1+\epsilon}$, $\epsilon > 0$ function on the support S_w (note that if Q is convex on an interval I, then it is automatically Lip 1 inside I). The sufficiency of the conjecture remains open for general convex Q's.

We have already mentioned, that if $w(x) = x^{\theta/(1-\theta)}$, then $S_w = [\theta^2, 1]$ ([23]). The generalization to Jacobi weights was done in [45], where it was shown that if $w(x) = (1 - x)^\alpha (1 + x)^\beta$, $\alpha, \beta \geq 0$, then the support of the extremal measure is

$$[\theta_2^2 - \theta_1^2 - \sqrt{\Delta}, \theta_2^2 - \theta_1^2 + \sqrt{\Delta}],$$

where $\theta_1 := \alpha/(1+\alpha+\beta)$, $\theta_2 := \beta/(1+\alpha+\beta)$ and

$$\Delta := \{1 - (\theta_1 + \theta_2)^2\}\{1 - (\theta_1 - \theta_2)^2\}.$$

In this case the approximation problem was settled by X. He and X. Li [13].

As a "midway" case between Jacobi weights and Freud-type exponential weights let us mention the Laguerre weights $w(x) = x^\alpha e^{-x}$, $\alpha \geq 0$, $\Sigma = [0, \infty)$, for which (see [38])

$$S_w = [1 + \alpha - \sqrt{(1 + \alpha^2) - \alpha^2}, 1 + \alpha + \sqrt{(1 + \alpha^2) - \alpha^2}].$$

In all these cases $Q = \log 1/w$ is convex, hence Theorem 4.3 can be applied and we can deduce the approximation property of the corresponding w on the interior of the support S_w. This is even true for the v. Golitschek–Saff–Varga theorem (note that in that theorem the function need not vanish at the right endpoint of $S_w = [\theta^2, 1]$), for it is enough to approximate $f \in C[0,1]$ that vanish on some $[0, \theta_1]$ with $\theta_1 > \theta^2$, and for such functions the claim follows from the fact that the function $F(x) = f(x\theta_1/\theta^2))$ can be uniformly approximated on $[0, \theta^2/\theta_1]$ by weighted polynomials of the form $x^{n\theta/(1-\theta)}P_n(x)$ (extend F to $[0,1]$ continuously so that it vanishes at 1, and apply Theorem 4.3 to F).

5   Preliminaries to the proofs

To prove Theorem 4.2 we shall need a few preliminary results.

First of all, let us note that properties (c) and (f) in Theorem A (see the introduction) imply the continuity of the equilibrium potential U^{μ_w} on its support S_w, and hence everywhere by Maria's theorem [51, Theorem III. 2]. Thus, U^{μ_w} is continuous everywhere.

We shall need to recognized the equilibrium measure in certain cases, and in such situations the following lemma comes in handy.

Lemma 5.1 *Let w be an admissible weight on Σ. If $\sigma \in \mathcal{M}(\Sigma)$ has compact support and finite logarithmic energy, and*

$$U^\sigma(z) + Q(z)$$

coincides with a constant F everywhere on the support of σ and is at least as large as F everywhere on Σ, then $\sigma = \mu_w$ and $F = F_w$. The same conclusion holds if the assumptions are true with the exception of finitely many points (or even if they are true only quasi–everywhere).

Proof. For the proof we recall the principle of domination for logarithmic potentials (see [17, Theorem 1.27]): *Let μ and ν be two positive finite Borel measures with compact support on $\mathbf{C}$, and suppose that the total mass of ν does not exceed that of μ. Assume further that μ has finite logarithmic energy. If, for some constant c, the inequality*

$$(5.1) \qquad\qquad U^\nu(z) + c \geq U^\mu(z)$$

holds μ-almost everywhere, then it holds for all $z \in \mathbf{C}$.

Theorem A (see the introduction) and the assumptions of the lemma imply that $U^{\mu_w}(z) \leq U^{\sigma}(z) - F + F_w$ for every $z \in \text{supp}(\mu_w)$, hence the principle of domination gives that the same inequality is true everywhere. Letting $z \to \infty$ we can conclude that $F \leq F_w$.

This argument can be repeated with μ_w and σ interchanged, hence we get that $F = F_w$, and then that the two potentials U^{μ_w} and U^{σ} concide everywhere. But then $\mu_w = \sigma$ (see e.g. [17, Theorem 1.12']).

The last statement in the lemma follows from the same considerations if we note that sets consisting of finitely many points must have zero μ_w- and $\sigma-$measures because these measures have finite logarithmic energy.
∎

Now we need the so called Fekete or Leja points (see [35]) associated with w. Let w be an admissible weight on the closed set $\Sigma \subseteq \mathbf{R}$. For an integer $n \geq 2$ we set

$$\delta_n^w := \sup_{z_1, \ldots, z_n \in \Sigma} \left\{ \prod_{1 \leq i < j \leq n} |z_i - z_j| w(z_i) w(z_j) \right\}^{2/n(n-1)}.$$

The supremum defining δ_n^w is obviously attained for some set

$$\mathcal{F}_n = \{z_1, \ldots, z_n\} \subseteq \Sigma.$$

These $\mathcal{F}_n$ are called n-th *Fekete sets* associated with w, or shortly *w-Fekete sets*. For fixed n, the sets $\mathcal{F}_n$ need not be unique; however in our consideration below we can use any choice of them. We shall need that the asymptotic distribution of Fekete points is the same as the equilibrium distribution μ_w: if

$$\nu_{\mathcal{F}_n}(A) := \frac{1}{n} \sum_{x \in \mathcal{F}_n \cap A} 1 = \frac{|\mathcal{F}_n \cap A|}{n},$$

where A is any Borel subset of $\mathbf{C}$ (i.e. $\nu_{\mathcal{F}_n}$ puts mass $1/n$ to every point $z_j \in \mathcal{F}_n$), then

(5.2) $$\lim_{n \to \infty} \nu_{\mathcal{F}_n} = \mu_w$$

in the weak* topology of measures (see [49, Lemma 2.2]).

With this property at hand we can easily verify that S_w is the smallest compact set S with the property that every weighted polynomial attains its norm on S (c.f. Theorem B in the introduction).

Lemma 5.2 *Let w be an admissible weight on Σ, and let $S \subseteq \Sigma$ be a closed set. If, for every $n = 1, 2, \ldots$ and every polynomial P_n with $\deg P_n \leq n$,*

$$\|w^n P_n\|_{\Sigma} = \|w^n P_n\|_{S},$$

then $S_w \subseteq S$.

Proof. Consider the Fekete sets $\mathcal{F}_n$ associated with w. We claim that for each n we can choose $\mathcal{F}_n \subseteq S$. This will prove the lemma because then the normalized counting measures $\nu_{\mathcal{F}_n}$ associated with $\mathcal{F}_n$ have support in S and converge in the weak* topology to the measure μ_w (see (5.2)), hence $S_w = \mathrm{supp}(\mu_w) \subseteq S$.

Let $\mathcal{F}_n = \{t_1, \ldots, t_n\}$ be any w-Fekete set, and suppose that $t_1, \ldots, t_{s-1} \in S$ but $t_s, \ldots, t_n \notin S$. Consider the polynomial

$$P_{n-1}(z) := \prod_{k \neq s} (z - t_k).$$

By the choice of w-Fekete points

$$\max_{z \in \Sigma} |P_{n-1}(z)| w(z)^{n-1} = |P_{n-1}(t_s)| w(t_s)^{n-1}.$$

By our assumption the weighted polynomial $w^{n-1} P_{n-1}$ attains its maximum modulus somewhere on S, hence there is a $t_s^* \in S$ with

$$\max_{z \in \Sigma} |P_{n-1}(z)| w(z)^{n-1} = |P_{n-1}(t_s^*)| w(t_s^*)^{n-1}.$$

Thus, together with $\{t_1, \ldots, t_{s-1}, t_s, t_{s+1}, \ldots, t_n\}$, the set $\{t_1, \ldots, t_{s-1}, t_s^*, t_{s+1}, \ldots, t_n\}$ will also be a w-Fekete set, and the latter set has already s points in S. We can continue this process and eventually arrive at a w-Fekete set contained in S. ∎

Now we are ready to prove a characterization of the points in the support that will be useful in our further considerations.

Lemma 5.3 *Let w be an admissible weight on Σ. Then $z \in \Sigma$ belongs to the support S_w of the extremal measure μ_w if and only if for every neighborhood B of z there exists a weighted polynomial $w^n P_n$, $\deg P_n \leq n$, such that $w^n |P_n|$ takes its maximum on Σ in $B \cap \Sigma$, and nowhere else.*

Proof. Let $z \in S_w$ and let B be any neighborhood of z. By applying Lemma 5.2 to the set $S = \Sigma \setminus B$, we get a weighted polynomial $P_n w^n$ with

$$\|w^n P_n\|_{\Sigma \setminus B} < \|w^n P_n\|_{\Sigma},$$

which shows that $w^n |P_n|$ takes its maximum on Σ in $B \cap \Sigma$, and nowhere else.

Conversely, if $w^n |P_n|$ takes its maximum on Σ only in $\Sigma \cap B$, then by Theorem B (see the introduction) we must have $B \cap S_w \neq \emptyset$, and of course if this is true for every neighborhood B of z, then $z \in S_w$. ∎

As an immediate consequence we obtain

Lemma 5.4 *If $\lambda > 1$, then*

$$S_{w^\lambda} \subseteq S_w,$$

i.e. the sets S_{w^λ} decrease as λ increases.

Proof. Let $z_0 \in S_{w^\lambda}$, and let B be any neighborhood of z_0. By Lemma 5.3 there are a natural number n, a polynomial P_n of degree at most n, a point $z_1 \in B \cap \Sigma$ and an $\eta > 0$ such that $|w^{n\lambda}(z_1)P_n(z_1)| = 1 + \eta$, but outside B we have $|w^{n\lambda}P_n| \leq 1$. Then w must be positive at z_1, say $w(z_1) \geq m > 0$. Furthermore, let M be an upper bound for w. We may assume $m < 1 < M$. For a positive integer l consider the polynomial P_n^l. We clearly have

$$(w(x))^{[ln\lambda]+1}|P_n(x)|^l \geq (1+\eta)^l m,$$

while for $x \in \Sigma \setminus B$

$$(w(z_1))^{[ln\lambda]+1}|P_n(z_1)|^l \leq M.$$

For large l this means that the weighted polynomial

$$w^{[ln\lambda]+1}P_n^l,$$

with $\deg P_n^l \leq nl \leq [ln\lambda] + 1$ takes its maximum modulus only in B. Since this is true for any neighborhood of z_0, we can infer from Lemma 5.3 that $z_0 \in S_w$, as we have claimed. ∎

Now we need the concept of balayage measure. Consider in $\mathbf{C}$ an open set G with compact boundary ∂G, and let μ be a measure with $\mathrm{supp}(\mu) \subseteq \overline{G}$. The problem of *balayage* (or "sweeping out") consists of finding a new measure $\overline{\mu}$ *supported on ∂G* such that $\|\overline{\mu}\| = \|\mu\|$ and

$$(5.3) \qquad U^\mu(z) = U^{\overline{\mu}}(z) \quad \text{for quasi-every} \quad z \notin G.$$

For bounded G such a measure always exists ([17, Chapter IV, §2/2]), but for unbounded ones (with $\mathrm{cap}(\partial G) > 0$) we have to replace (5.3) by

$$(5.4) \qquad U^\mu(z) = U^{\overline{\mu}}(z) + c \quad \text{for quasi-every} \quad z \notin G.$$

with some constant c.

Besides (5.3)–(5.4) we also know that

$$(5.5) \qquad U^{\overline{\mu}}(z) \leq U^\mu(z)$$

respectively
$$(5.6) \qquad U^{\overline{\mu}}(z) \leq U^\mu(z) + c$$

holds for all $z \in \mathbf{C}$. Furthermore, if G is regular with respect to the Dirichlet problem, then we have equality in (5.3) and (5.4). This is the case for example when G is the complement of finitely many closed intervals on the real line.

Very often we have to take the balayage of μ out of G even if its support is not contained in $\overline{G}$. In that case we take the balayage of the restriction of μ to G onto ∂G, and leave the part of μ lying outside G unchanged.

Now the first part of the following lemma is an immediate consequence of these properties, Theorem A and Lemma 5.1. The second part follows from Lemma 5.3.

Lemma 5.5 *Let w be admissible, and $K \subseteq S_w$ a regular compact subset of S_w. Then*

$$\mu_w\big|_K = \overline{\mu_w},$$

where ‾ *indicates taking balayage onto K out of $C \backslash K$. In particular, $S_w\big|_K = K$.*

The lemma is true without the regularity assumption but we will not need this stronger statement. Recall that K is called regular if $C \backslash K$ is regular with respect to the Dirichlet problem. Let us also mention that regularity is characterized by the Wiener condition (1.2).

The next lemma can be esasily proved by the method of Lemma 5.1 if we use the properties (1.7)–(1.8) of the equilibrium measures.

Lemma 5.6 *If $\lambda > 1$ and $S_{w^\lambda} = S_w$, furthermore this is a regular set, then*

$$\mu_w = \frac{1}{\lambda}\mu_{w^\lambda} + \left(1 - \frac{1}{\lambda}\right)\omega_{S_w}.$$

Again, here the regularity can be dropped, but we shall anyway need the lemma only in the case when the support in question is an interval.

In what follows, let v_w denote the density function of μ_w (wherever it exists).

If the two supports S_w and S_{w^λ}, $\lambda > 1$ are not the same, then we only have inequalities for the corresponding extremal measures.

Lemma 5.7 *If $\lambda > 1$, then*

$$\mu_w\big|_{S_{w^\lambda}} \le \frac{1}{\lambda}\mu_{w^\lambda} + \left(1 - \frac{1}{\lambda}\right)\omega_{S_{w^\lambda}}$$

and

$$\mu_w\big|_{S_w} \ge \frac{1}{\lambda}\mu_{w^\lambda} + \left(1 - \frac{1}{\lambda}\right)\omega_{S_w}\big|_{S_w^\lambda}.$$

Proof. The proof is based on the following theorem of de la Vallée Poussin from [5] (see also [4, 11, 7], and also [50]): *Let μ and ν be two measures of compact support, and let Ω be a domain in which both potentials U^μ and U^ν are finite and satisfy with some constant c the inequality*

$$(5.7) \qquad\qquad U^\mu(z) \le U^\nu(z) + c, \qquad z \in \Omega.$$

If A is the subset of Ω in which equality holds in (5.7), then $\nu\big|_A \le \mu\big|_A$, i.e. for every Borel subset B of A the inequality $\nu(B) \le \mu(B)$ holds.

Consider the two potentials corresponding to the two measures μ_w and μ_{w^λ} with some $\lambda > 1$. It follows from Theorem A (see the introduction) and (1.8) that with $\omega = \omega_{S_w}$ (recall that this is the equilibrium measure of the set S_w)

$$\frac{1}{\lambda}\left(U^{\mu_{w^\lambda}}(z) - F_{w^\lambda}\right) + \left(1 - \frac{1}{\lambda}\right)\left(U^\omega(z) - \log\frac{1}{\mathrm{cap}(S_w)}\right) \ge U^{\mu_w}(z) - F_w$$

for quasi–every $z \in S_w$, hence by the principle of domination (see the proof of Lemma 5.1) we have this inequality everywhere (recall that if a measure μ has finite logarithmic energy, then every set of zero capacity has zero μ–measure). Furthermore, Theorem A,(f), (1.8) and Lemma 5.4 imply that the equality sign holds for quasi–every $z \in S_{w^\lambda}$. Now each of the measures μ_w, μ_{w^λ} and ω have finite logarithmic energy, hence they vanish on sets of zero capacity, thus by applying the above theorem of de la Vallée Poussin we can conclude the second inequality:

$$\mu_w\big|_{S_{w^\lambda}} \geq \frac{1}{\lambda}\mu_{w^\lambda} + \left(1 - \frac{1}{\lambda}\right)\omega_{S_w}\big|_{S_w^\lambda}.$$

The first one can be shown with the same argument if we notice that with $\overline{\omega} = \omega_{S_{w^\lambda}}$ we have

$$\frac{1}{\lambda}\left(U^{\mu_{w^\lambda}}(z) - F_{w^\lambda}\right) + \left(1 - \frac{1}{\lambda}\right)\left(U^{\overline{\omega}}(z) - \log\frac{1}{\mathrm{cap}(S_{w^\lambda})}\right) \leq U^{\mu_w}(z) - F_w$$

for every $z \in S_{w^\lambda}$, hence by the principle of domination we have this inequality everywhere. Furthermore, by Theorem A,(f) and Lemma 5.4 the equality sign holds for quasi–every $z \in S_{w^\lambda}$. Thus, the first inequality follows as before from the aforementioned theorem of de la Vallée Poussin. ∎

In order to apply the preceding lemma we shall need a convenient criterion for concluding that a point x_0 from S_w belongs to some S_{w^λ}, $\lambda > 1$, as well (recall Lemma 5.4 that these support are decreasing).

Lemma 5.8 *Suppose x_0 is a point in the interior of S_w, v_w is continuous in a neighborhood of x_0, and $v_w(t) > \epsilon_0$ for $|t - x_0| \leq \epsilon_1$ for some $\epsilon_0 > 0$ and $\epsilon_1 > 0$. Then for $\lambda \leq 1/(1 - \epsilon_0\epsilon_1)$ the point x_0 is in the interior of S_{w^λ}, furthermore, v_{w^λ} is also continuous in a neighborhood of x_0.*

Proof. We begin the proof by the following observation. If for w we consider minimizing the weighted energy (1.5) on Σ and on some closed set $S_w \subseteq \Sigma_1 \subseteq \Sigma$, then we arrive at the same extremal measure S_w. Seeing that $S_{w^\lambda} \subseteq S_w$ (Lemma 5.4), this shows that in the proof we may assume without loss of generality that $\Sigma = S_w$.

Let ν_0 be the measure the density of which is ϵ_0 on $[x_0 - \epsilon_1/2, x_0 + \epsilon_1/2]$ and 0 otherwise, and consider the positive measure

$$\nu_1 = \frac{1}{1 - \epsilon_0\epsilon_1}(\mu_w - \nu_0)$$

of total mass 1, and the weight function

$$w_1(x) = \exp(U^{\nu_1}(x))$$

that it generates. By Lemma 5.1 the extremal measure corresponding to w_1 coincides with ν_1, and so $x_0 \in S_{w_1}$. Hence (see Lemma 5.3) if B is a neighborhood of x_0, then there is a polynomial P_n such that $w_1^n|P_n|$ attains its maximum in $B \cap \Sigma$ and nowhere else on Σ.

The potential of the measure

$$\frac{1}{1 - \epsilon_0\epsilon_1}\nu_0$$

is symmetric about x_0, attains its maximum at x_0 and decreases to the right and increases to the left of x_0. But then for the weight

$$w_2(x) = w_1(x)\exp(U^{\nu_0/(1-\epsilon_0\epsilon_1)}(x))$$

the weighted polynomial $w_1^n P_n$ can also attain its maximum only in B. Since this can be done for every neighborhood B of x_0, it follows that $x_0 \in S_{w_2}$. However, the weight function

$$w_2(x) = \exp(U^{\mu_w}(x)/(1 - \epsilon_0\epsilon_1))$$

and w^λ with $\lambda = 1/(1 - \epsilon_0\epsilon_1)$ differ on S_w only in a multiplicative constant, which, together with the relation $\Sigma = S_w$, means that $\mu_{w^\lambda} = \mu_{w_2}$, and so $S_{w^\lambda} = S_{w_2}$. Thus, $x_0 \in S_{w^\lambda}$ as is claimed in the lemma (see also Lemma 5.4). Furthermore, the same proof can be carried out with x_1 in place of x_0 for every $x_1 \in [x_0 - \epsilon_1/2, x_0 + \epsilon_1/2]$ lying sufficiently close to x_0, hence x_0 is actually in the interior of S_{w^λ}, and $\mu)w^\lambda$ has a continuous density in a neighborhood of x_0.

It has left to show the continuity of the density function v_{w^λ} at x_0. Let $I = [x_0 - \epsilon_1/2, x_0 + \epsilon_1/2]$, and let $\bar{}$ denote taking balayage onto I out of $\mathbf{C} \setminus I$. By Lemmas 5.5 and 5.6 we have

$$\mu_w\big|_I = \overline{\mu_w}, \qquad \mu_{w^\lambda}\big|_I = \overline{\mu_{w^\lambda}}$$

and

$$\mu_w\big|_I = \frac{1}{\lambda}\mu_{w^\lambda}\big|_I + \left(1 - \frac{1}{\lambda}\right)\omega_I,$$

from which we get the formula

$$\begin{aligned}
\mu_{w^\lambda}\big|_I &= \overline{\mu_{w^\lambda}} - \overline{\mu_{w^\lambda}\big|(\mathbf{R}\setminus I)} = \lambda\mu_w\big|_I - (\lambda-1)\omega_I - \overline{\mu_{w^\lambda}\big|(\mathbf{R}\setminus I)} \\
&= \lambda\mu_w\big|_I + \lambda\overline{\mu_w}\big|(\mathbf{R}\setminus I) - (\lambda-1)\omega_I - \overline{\mu_{w^\lambda}\big|(\mathbf{R}\setminus I)}.
\end{aligned}$$

Now on the right each term beginning with the second one has continuous density in the interior of I, furthermore, by our assumption the first term also has continuous density at x_0, and these prove that μ_{w^λ} has continuous density function at x_0.

6   Proof of Theorems 4.1, 4.2 and 4.3

After the preliminaries of the preceding section we are in the position to prove Theorems 4.1–4.3.

Proof of Theorem 4.1.   Let $\mathcal{N}_1$ be the set of the even natural numbers. By considering $(w^n P_n)^2$ we can see that for $n \in \mathcal{N}_1$ there are polynomials R_n of degree at most n such that $w^n R_n$ converges to f^2 on $\mathcal{S}_w \cup \{x_0\}$, and here f^2 is already nonnegative. Let the minimum and maximum of f^2 on $\mathcal{S}_w$ be m and M, respectively.

Let us suppose on the contrary that $\alpha := f^2(x_0) > 0$, which will lead us to a contradiction as follows. We distinguish three cases.

Case I. f^2 is not constant on $\mathcal{S}_w$ (i.e. $0 \le m < M$). We shall utilize the existence of a polynomial $U(x)$ *without constant term* such that $U(x) \ge 0$ on $[0, M] \cup \{\alpha\}$, $U(m) \ne U(M)$ and on $[0, M] \cup \{\alpha\}$ the polynomial U takes its maximum at the point α. Such U's can be constructed as follows. Let $K = \max\{2M, \alpha\}$. If we choose the polynomial $U^*(x)$ as $x(K - x)^s$, then U^* increases on the interval $[0, K/(s+1)]$ and decreases on $[K/(s+1), K]$, hence by selecting s bigger than K/α and setting $U(x) = U^*(Kx/(s+1)\alpha)$, the polynomial U will be nonnegative on $[0, K]$, will increase on $[0, \alpha]$ and decrease on $[\alpha, K]$. If $U(m) \ne U(M)$, then we are ready. In the opposite case certainly $m \ne 0$, and we can set $U(x) = U^{**}(x)U^*(Kx/(s+1)\alpha)$ with a polynomial U^{**} that is nonnegative on $[0, K]$, takes different values at m and M and takes its maximum on $[0, K]$ at the point α.

We claim that there is a subsequence $\mathcal{N}_2 \subseteq \mathcal{N}_1$ and for each $n \in \mathcal{N}_2$ a polynomial S_n of degree at most n such that $w^n S_n$ uniformly tends to $g := U(f^2)$ on $\mathcal{S}_w \cup \{x_0\}$ as $n \to \infty$, $n \in \mathcal{N}_2$. In fact, let k be the degree of U. Since $w^{2n} R_{2n} \to f^2$, $n \to \infty$, we can see that for any $j \le k$ the weighted polynomials $w^{2k!n} R^j_{2k!n/j}$ converge to $(f^2)^j$. Thus, if $U(x) = \sum_{j=1}^{k} a_j x^j$, then the polynomials

$$S_{2k!n}(x) = \sum_{j=1}^{k} a_j R^j_{2k!n/j}(x)$$

of degree at most $2k!n$, will do the job (thus, we can choose $\mathcal{N}_2 = \{2k!n \mid n = 1, 2 \ldots\}$).

By squaring again and considering g^2 instead of $g := U(f^2)$ if necessary, we can also suppose that the polynomials S_n are nonnegative. Furthermore, the degree of S_n can be exactly n, for we can always add factors of the form $(L - x)^2/L^2$ to S_n with sufficiently large L and these factors do not to change the properties of S_n on the compact set $\mathcal{S}_w \cup \{x_0\}$. In a similar manner, without loss of generality we may suppose that S_n does not vanish on the real line, for we can always replace any double real zero a of S_n by two conjugate complex zeros lying close to a. Note also that g is not constant on $\mathcal{S}_w$ because we had $U(m) \ne U(M)$ for the polynomial U.

If we use Theorem A from the introduction we can see that

$$(6.1) \qquad n(U^{\mu_w}(x) - F_w) - \log \frac{1}{S_n(x)} \to \log g(x)$$

uniformly for all $x \in \mathcal{S}_w \cup \{x_0\}$ as $n \to \infty$, $n \in \mathcal{N}_2$. If M_1 denotes the maximum of g on $\mathcal{S}_w$ then it follows that there is a positive sequence $\{\epsilon_n\}$ converging to 0 such that

$$(6.2) \qquad - n(U^{\mu_w}(x) - F_w) + \log \frac{1}{S_n(x)} + \log M_1 + \epsilon_n \geq 0, \qquad x \in \mathcal{S}_w.$$

Recall now the choice of U, according to which the function g attains its maximum M_1 at the point x_0 on $\mathcal{S}_w \cup \{x_0\}$. Thus,

$$(6.3) \qquad - n(U^{\mu_w}(x_0) - F_w) + \log \frac{1}{S_n(x_0)} + \log M_1 + \epsilon_n \to 0$$

as $n \to \infty$, $n \in \mathcal{N}_2$.

Let ν_n be the counting measure on the set of zeros of S_n (counting multiplicity) and let $\overline{\nu_n}$ be its balayage out of $\overline{\mathbf{C}} \setminus \mathcal{S}_w$ onto $\mathcal{S}_w$ with balayage constant d_n, i.e. (see (5.4) in Section 5)

$$U^{\overline{\nu_n}}(x) = U^{\nu_n}(x) + d_n = \log \frac{1}{S_n(x)} + d_n$$

for quasi-every $x \in \mathcal{S}_w$. Then (6.2) goes into

$$(6.4) \qquad - n(U^{\mu_w}(x) - F_w) + U^{\overline{\nu_n}}(x) - d_n + \log M_1 + \epsilon_n \geq 0$$

for quasi-every $x \in \mathcal{S}_w$, hence by the principle of domination (see the proof of Lemma 5.1) this inequality holds true *for all* $x \in \overline{\mathbf{C}}$. On the other hand, from the nonincreasing character of taking balayage (see (5.6)) we can conclude from (6.3) that

$$(6.5) \quad \limsup_{n \to \infty,\ n \in \mathcal{N}_2} \left(-n(U^{\mu_w}(x_0) - F_w) + U^{\overline{\nu_n}}(x_0) - d_n + \log M_1 + \epsilon_n \right) \leq 0,$$

hence we must have here equality in the limit. But the function on the left is harmonic and nonnegative on $\overline{\mathbf{C}} \setminus \mathcal{S}_w$, so Harnack's theorem implies that

$$\lim_{n \to \infty,\ n \in \mathcal{N}_2} \left(-n(U^{\mu_w}(z) - F_w) + U^{\overline{\nu_n}}(z) - d_n + \log M_1 \right) = 0$$

uniformly on compact subsets of $\overline{\mathbf{C}} \setminus \mathcal{S}_w$. In particular, for $z = \infty$ we obtain

$$(6.6) \qquad \lim_{n \to \infty,\ n \in \mathcal{N}_2} (nF_w - d_n + \log M_1) = 0,$$

where we used that the total mass of the measures $n\mu_w$ and $\overline{\nu_n}$ are both n.

(6.1) yields with some monotone decreasing sequence $\{\eta_n\}$ tending to zero that

$$-n(U^{\mu_w}(x) - F_w) + U^{\overline{\nu_n}}(x) - d_n \geq -\log(g(x) + \eta_n)$$

for quasi–every $x \in S_w$, which combined with the preceding limit relation gives
for $n \in \mathcal{N}_2$

$$(6.7) \qquad -nU^{\mu_w}(x) + U^{\overline{\nu_n}}(x) \geq \log \frac{M_1}{g(x) + \eta_n} - \rho_n$$

for quasi–every $x \in S_w$, with some sequence $\{\rho_n\}$ tending to zero.

Since the equilibrium measure $\omega = \omega_{S_w}$ of S_w has finite logarithmic energy,
sets of zero capacity must have zero ω–measure, hence the preceding inequality
holds true ω–almost everywhere. Thus, we can integrate this inequality with
respect to ω. Using the fact that the measures $n\mu_w$ and $\overline{\nu_n}$ have finite logarithmic energy (recall that S_n did not have real zero, hence the balayage measure
of its zero counting measure is of finite logarithmic energy) and that we have

$$U^\omega(x) = \log \frac{1}{\mathrm{cap}(S_w)}$$

for quasi–every $x \in S_w$ (see (1.8)) we get from Fubini's theorem that

$$\int nU^{\mu_w} d\omega = n \int U^\omega d\mu_w = n \log \frac{1}{\mathrm{cap}(S_w)} = \int U^\omega d\overline{\nu_n} = \int U^{\overline{\nu_n}} d\omega,$$

and so during the integration of (6.7) against ω the left hand side becomes zero.
Thus, in the limit we get from the monoton convergence theorem that

$$\int \log \frac{M_1}{g(x)} d\omega(x) = 0,$$

which can happen only if $g(x) = M_1$ ω–almost everywhere (recall that M_1 was
the maximum of g on S_w). But this is certainly not the case for g is not constant
on S_w, so in some neighborhood O of an $x_1 \in S_w$ we have $g < M_1$, and since
$\mathrm{cap}(O \cap S_w) > 0$ (recall that S_w was the support of the measure μ_w of finite
logarithmic energy), we must have $\omega(O \cap S_w) > 0$. The obtained contradiction
proves the claim.

Case II. f^2 is constant on S_w, but w is not. This case can be easily reduced to
the previous one, for if $w^{2n} R_{2n}$ tends to a constant on S_w, then $w^{2n+2} R_{2n}$ will
tend to w^2 which is not constant and $\deg R_{2n} \leq 2n + 2$.

Case III. Both f^2 and w are constant on S_w (in this case $\mu_w = \omega_{S_w}$). We
can assume by proper normalization that $w(x) \equiv f^2(x) \equiv 1$ on S_w. By the
maximum principle we have strict inequality in

$$U^\omega(z) \leq \log \frac{1}{\mathrm{cap}(S_w)}$$

when $z \notin S_w$ (see (1.7)), and since in the present case

$$w(x_0) \leq \exp \left(U^\omega(x_0) - \log \frac{1}{\mathrm{cap}(S_w)} \right),$$

(see Theorem A,(d) in the introduction), we can conclude that $w(x_0) < 1$. Hence by considering $w^{2n+2}R_{2n}$ instead of $w^{2n}R_{2n}$ if necessary, we can assume that $f^2(x_0) =: \alpha \neq 1$. Choose now a polynomial U without constant term so that $U(1) < 1 < 3 < U(\alpha)$ is satisfied. Then, as before, there are some polynomials S_n, $n \in \mathcal{N}_2$ such that $w^n S_n \to U(f^2)$ uniformly on $\mathcal{S}_w \cup \{x_0\}$, i.e. $w^n(x)S_n(x)$ uniformly converges to $U(1) < 1$ on $\mathcal{S}_w$ while it converges to $U(\alpha) > 3$ at x_0. This would mean that the sup norm of $w^n S_n$ on $\mathcal{S}_w$ is smaller than 2 for all large $n \in \mathcal{N}_2$, and at the same time these weighted polynomials take values bigger than 2 at x_0 — a contradiction to Theorem B in the introduction. ∎

Proof of Theorem 4.2.   Let $f \in C_0(S^w)$. We follow the proof of Theorem 1.1. In the present case J_ϵ denotes the set of points $x \in S^w$ which are of distance $\geq \epsilon$ form the complement $\mathbf{R} \setminus S^w$.

Let J^* be an arbitrary finite interval. Eventually we will choose J^* so that in $\Sigma \setminus J^*$ we have

(6.8) $$U^{\mu_w}(z) \geq -Q(z) + F_w + 1$$

(c.f. Theorem A and the definition of the admissibility of w in the introduction), but for secure a free choice of J^* later, it can be arbitrary at this point.

First we show that it is enough to verify the following analogues of (2.1) and (2.2) for arbitrary J^*:

(6.9) $$w^n(x)|Q_n(x)| = \exp(g_L(x) + R_L(x)), \qquad x \in J_\epsilon,$$

where the remainder term $R_L(x)$ satisfies $|R_L(x)| \leq C_\epsilon/L$ uniformly in $x \in J_\epsilon$ with some $C_\epsilon \geq 1$, and uniformly in $x \in \Sigma \cap J^*$

(6.10) $$w^n(x)|Q_n(x)| = e^{o(n)}.$$

In fact, suppose this is true, and exactly as in Section 2 we apply it to w^λ instead of w with some $\lambda > 1$. We can do this, because by Lemma 5.8 there is a $\lambda > 1$ such that the set $J_{\epsilon/2}$ is in the support S_{w^λ} of μ_{w^λ} and it has continuous density there, furthermore, exactly as U^{μ_w} the potential $U^{\mu_{w^\lambda}}$ is continuous everywhere; and these are the only properties that we shall use in deriving (6.9) and (6.10) below. Hence, by chosing $\lambda > 1$ close to 1 we get that there are polynomials $Q_{[n/\lambda]}$ of degree at most $[n/\lambda]$ such that with some g_L and R_L as above

$$w^n(x)|Q_{[n/\lambda]}(x)| = \exp(g_L(x) - (n - \lambda[n/\lambda])Q(x) + R_L(x)), \qquad x \in J_\epsilon$$

and

(6.11) $$w^n(x)|Q_{[n/\lambda]}(x)| = e^{o(n)}, \qquad x \in \Sigma \cap J^*$$

where now J^* is an interval satisfying (6.8).

Let $1/\lambda < \tau < 1$. We consider the polynomials $S_{n-[n/\lambda]}$ from the proof of Theorem 1.1, except that now we request that their degree be at most $[(\tau -$

$1/\lambda)n]$ (hence we write $S_{[(\tau-1/\lambda)n]}$ for them), and we need their third property (2.3) only in the form

(6.12) $$|S_{[(\tau-1/\lambda)n]}(x)| \le 1, \qquad x \in J^* \setminus J_\epsilon.$$

Since the disjoint sets $J^* \setminus J_\epsilon$ and $J_{2\epsilon}$ consist of finitely many intervals, there is a $0 < c < 1$ and for each m polynomials R_m of degree at most m such that

(6.13) $$|R_m(x) - 1| \le c^m \qquad \text{for } x \in J_{2\epsilon},$$

(6.14) $$|R_m(x)| \le c^m \qquad \text{for } x \in J^* \setminus J_\epsilon$$

and
(6.15) $$0 \le R_m(x) \le 1 \qquad \text{for } x \in J_\epsilon \setminus J_{2\epsilon}$$

(see e.g. [15, Theorem 3] where such polynomials were constructed for two disjoint intervals, from which the R_m's with the stated properties can be easily patched together).

Finally, we set

$$P_n(x) = Q_{[n/\lambda]}(x)S_{[(\tau-1/\lambda)n]}(x)R_{[(1-\tau)n]}(x),$$

which has degree at most n. Exactly as in Section 2 we get that if $\eta > 0$, then by choosing $\epsilon > 0$, $\lambda > 1$ and L appropriately, for sufficiently large n the difference $|w^n|P_n| - f|$ will be smaller than 3η on the set $J^* \cap \Sigma$. The only remark we have to make is that by (6.11), (6.12) and (6.14), the weighted polynomial $w^n P_n$ is exponentially small on $(\Sigma \cap J^*) \setminus J_\epsilon$. But by (6.8) and Theorem B (see the introduction) the same is true on $\Sigma \setminus J^*$, and this proves that

$$|w^n|P_n| - f| \le 3\eta$$

for every $x \in \Sigma$ provided n is sufficiently large.

Thus, it has left to prove (6.9) and (6.10). The proof of (6.9) is identical to the proof of (2.1) given in Section 2, there is no need to change anything. As for (6.10), with the measure μ_n from the proof of Theorem 1.1 we get by the continuity of the potential U^{μ_w} that

$$|U^{\mu_w}(x - iL/n) - U^{\mu_w}(x)| = o(1)$$

as $n \to \infty$ uniformly in $x \in \mathbf{R}$, while the estimate

$$\log |Q_n(x)| + nU^{\mu_n}(x) \le 3 \log 2Dn/L$$

with D equal to the diameter of J^* follows exactly as (2.9). These and

$$w(x) \le \exp(U^{\mu_w}(x) - F_w)$$

(see Theorem A,(d) in the introduction) prove (6.10), and the proof is complete. ∎

Proof of Theorem 4.3.   We show that Theorem 4.3 is a consequence of Theorem 4.2.

Let $\Sigma = \cup I_j$, where on the right we have a finite and disjoint union, and suppose that Q is convex on each of the I_j's. It was proved in [34, Theorem 2.2] that then S_w consists of finitely many intervals at most one lying in any of the I_j's.

Let I be any subinterval of the interior of S_w. We recall Lemma 5.5 according to which

$$\mu_w\big|_I = \overline{\mu_w}\big|_I - \overline{\mu_w}\big|(\mathbf{R} \setminus I),$$

where $^-$ indicates taking balayage onto I out of $\mathbf{C} \setminus I$. The second measure on the right has continuous density inside I, and the same is true for the first one by [49, Lemma 4.5] (c.f. also the proof of Theorem 8.3) because the support of $\mu_w\big|_I$ is I, and $w\big|_I$ is a $C^{1+\epsilon}$–function on the support. Thus, we can conclude that μ_w has continuous density in the interior of S_w.

In order to be able to apply Theorem 4.2 we have to show that the density v_w of the extremal measure μ_w cannot vanish at any interior point of S_w. But this is an immediate consequence of Lemma 5.7 according to which we have

$$\mu_w\big|_{S_{w^\lambda}} \geq \frac{1}{\lambda}\mu_{w^\lambda} + \left(1 - \frac{1}{\lambda}\right)\omega_{S_w}\big|_{S_w^\lambda}$$

and this clearly rules out that the density of μ_w vanishes at a point x_0 unless x_0 does not belong to the interior of S_{w^λ}. Thus, if $x_0 \in S_w$ does belong to the interior of some S_{w^λ}, then at x_0 the measure μ_w has positive density. But every interior point x_0 of S_w must belong to the interior of at least one S_{w^λ}. In fact, since every S_{w^λ} consists of intervals at most one of which can lie in any I_j, it is enough to prove that in any neighborhood of any point x_0 of S_w there is a point x_1 lying in some S_{w^λ}, because then this and the decreasing character of the supports S_{w^λ} (Lemma 5.4) imply our claim concerning every point in $\text{Int}(S_w)$ lying in the interior of some S_{w^λ}. But if $x_0 \in S_w$, and B is any neighborhood of x_0, then there are an n and a polynomial P_n of degree at most n such that $w^n|P_n|$ attains its maximum in Σ at some point of $B \cap \Sigma$ and nowhere outside of B. By continuity then the same is true of $w^{\lambda n}|P_n|$ with some $\lambda > 1$ sufficiently close to 1, and so Theorem B implies that $B \cap S_{w^\lambda} \neq \emptyset$. With this the proof is complete.   ∎

As for Theorem 4.4, compactness shows that it is enough to consider functions that vanish outside some $\text{Int}(S_{w^\lambda})$, and for such functions the preceding proof gives the appropriate approximation.

7   Construction of Examples 4.5 and 4.6

In this section we construct the two theoretically important Examples 4.5 and 4.6.

7.1   Example 4.5

In Example 4.5 we have to construct a weight w such that the support of the corresponding extremal measure is $[-1, 1]$, this measure has continuous density in $(-1, 1)$ which is positive everywhere except at 0, and still no function that is nonzero at 0 is the uniform limit of weighted polynomials. Hence, in this case the largest set for the approximation problem of Section 4.1 is the restricted support $(-1, 0) \cup (0, 1)$.

The idea of the proof is to construct a density function v that is positive and continuous on $(-1, 0) \cup (0, 1)$ and tends to zero very fast at the origin.

We shall construct an even w by inductively choosing density functions that will converge in some sense to the density of w. In fact, for each $k \geq 2$ we shall construct a function v_k with the following properties: Let $S_k = [-1 + 2^{-k}, -2^{-k}] \cup [2^{-k}, 1 - 2^{-k}]$. For every $k \geq 2$

1. v_k is even, positive and continuous on S_k and zero elsewhere,

2. the integral of v_k is 1, i.e. v_k generates an absolutely continuous positive measure of unit mass with support on S_k,

3. $v_k(x) \leq 1/k$ if $|x| \leq 2^{1-k}$, furthermore $v_k(\pm 2^{-k}) \leq 1/(k+1)$,

4. $|v_k(x) - v_{k-1}(x)| \leq 2^{-k}$ if $x \in S_{k-1}$,

5. there exists an $n_k > k$ such that if $w_k(x) = \exp(U^{v_k}(x))$, where U^{v_k} denotes the potential associated with the measure generated by v_k, then for every polynomial P_{n_k} of degree at most n_k the inequality
    $$w_k^{n_k}(x)|P_{n_k}(x)| \leq 1, \qquad x \in S_k,$$
    implies
    $$w_k^{n_k}(0)|P_{n_k}(0)| < \frac{1}{k}.$$

From here the construction of w with the desired properties will be simple, so first let us consider the construction of v_k. For $k = 1$ define v_k to satisfy the first three conditions, and now let us proceed with the construction of v_{k+1} provided v_k is already known.

Let
$$I_k = [-1 + 2^{-k-1}, -1 + 3 \cdot 2^{-k-2}] \cup [1 - 3 \cdot 2^{-k-2}, 1 - 2^{-k-1}].$$

Note that $I_k \subset S_{k+1}$ is disjoint from S_k, hence if ω_{S_k} is the equilibrium measure associated with S_k (see Section 4.1), then

$$(7.1) \qquad \log \frac{1}{\mathrm{cap}(S_k)} - \max_{x \in I_k} U^{\omega_{S_k}}(x) =: \eta_k > 0$$

(see (1.7)–(1.8) and apply the maximum principle according to which the potential of ω_{S_k} is strictly smaller than $\log(1/\mathrm{cap}(S_k))$ outside S_k).

Choose now a $\beta = \beta_k > 1$ arbitrarily. We claim

Lemma 7.1 *There is a sequence θ_n tending to 0 such that if P_n is an arbitrary polynomial of degree at most n, and*

$$(7.2) \qquad w_k^n(x)|P_n(x)| \le 1, \qquad x \in S_k,$$

and

$$(7.3) \qquad w_k^n(x)|P_n(x)|\beta^n \le 1, \qquad x \in I_k,$$

then

$$(7.4) \qquad w_k^n(0)|P_n(0)| \le \theta_n.$$

Recall that $w_k(x) = \exp(U^{v_k}(x))$.

Proof of of Lemma 7.1. We begin the proof by reducing it to the case when P_n has a special form. First of all by considering $(P_n(x) + P_n(-x))/2$ instead of P_n we can assume that P_n is even. Let

$$P_n(x) = a_n \prod_j (x^2 - \alpha_j),$$

and

$$P_n^*(x) = |a_n| \prod_j (x^2 - |\alpha_j|).$$

Since $|P_n(0)| = |P_n^*(0)|$, and for every $x \in \mathbf{R}$ the inequality $|P_n(x)| \ge |P_n^*(x)|$ holds because of the obvious inequality $|x^2 - \alpha_j| \ge |x^2 - |\alpha_j||$, without loss of generality we can assume $P_n = P_n^*$, i.e. that the leading coefficient a_n is positive, and $\alpha_j \ge 0$ for all j, which also means that P_n has only real zeros.

Let

$$0 \le \alpha_1, \dots, \alpha_s < 2^{-2k} \quad \text{and} \quad \alpha_{s+1}, \dots \ge 2^{-2k},$$

and now consider

$$P_n^{**}(x) := P_n(x) \left(\prod_{j=1}^s \frac{x^2 - 2^{-2k}}{x^2 - \alpha_j} \cdot \frac{\alpha_j}{2^{-2k}} \right).$$

Again $P_n(0) = P_n^{**}(0)$, and for every $|x| \ge 2^{-k}$ the inequality $|P_n(x)| \ge |P_n^{**}(x)|$ holds because for such x the function $|x^2 - \alpha|/\alpha$ decreases in α on the interval $[0, 2^{-2k})$. Thus, (7.2)–(7.3) hold also with P_n replaced by P_n^{**}, while the status of the inequality (7.4) remains unchanged, hence without loss of genrality we can assume $P_n = P_n^{**}$, i.e. that P_n does not have any zero on the interval $(-2^{-k}, 2^{-k})$.

Let M be arbitrary large. We distinguish two cases.

Case 1. There are at least M zeros of P_n in I_k. If we set

$$U_n(x) = \log \frac{1}{|P_n(x)|} - nU^{v_k}(x),$$

then (7.2) takes the form

(7.5) $$U_n(x) \geq 0$$

for every $x \in S_k = \mathrm{supp}(v_k)$. We can apply the principle of domination (see the proof of Lemma 5.1 in Section 5) to deduce (7.5) for all $x \in \mathbf{C}$. Now let us recall Harnack's principle, according to which if U is nonnegative and harmonic on a domain D, and $y_1, y_2 \in D$, then there exists a K independent of U such that $U(y_1) \leq KU(y_2)$. We set $D = (\mathbf{C} \setminus \mathbf{R}) \cup (-2^{-k}, 2^{-k})$. Since all the zeros of P_n lie in $\mathbf{R} \setminus (-2^{-k}, 2^{-k})$, we can conclude the harmonicity of U_n in D, and so for $y_1 = 0$, $y_2 = \infty$ there is a K such that

(7.6) $$U_n(\infty) \leq KU_n(0)$$

independently of n.

Now integrate (7.5) with respect to the equilibrium measure ω_{S_k} of S_k. Taking into account (1.7)–(1.8) we get from Fubini's theorem with the counting measure ν_n on the zeros of P_n

$$
\begin{aligned}
n \log \frac{1}{\mathrm{cap}(S_k)} &= n \int U^{\omega_{S_k}} v_k = \int nU^{v_k} d\omega_{S_k} \leq \int \log \frac{1}{|P_n|} d\omega_{S_k} \\
&= \log \frac{1}{a_n} + \int U^{\nu_n} d\omega_{S_k} = \log \frac{1}{a_n} + \int U^{\omega_{S_k}} d\nu_n \\
&\leq \log \frac{1}{a_n} + n \log \frac{1}{\mathrm{cap}(S_k)} - M\eta_k,
\end{aligned}
$$

where, at the very last step we used (7.1) and the fact that at least M of the zeros of P_n lie on I_k. This implies that

(7.7) $$U_n(\infty) = \log \frac{1}{a_n} \geq M\eta_k,$$

and so, in view of (7.6), we can conclude

$$U_n(0) \geq M \frac{\eta_k}{K}.$$

Case 2. There are at most $M - 1$ zeros of P_n in I_k. Let $J_1, \ldots, J_M$ be disjoint open subintervals of I_k and $t_1, \ldots, t_M$ one-one point in each. Choose connected open sets D_j, $j = 1, \ldots, M$ containing the points 0 and t_j such that

$$D_j \cap \mathbf{R} \subseteq J_j \cup (-2^{-k}, 2^{-k}).$$

By Harnack's inequality there are numbers K_j such that if U is a nonnegative harmonic function on D_j, then

$$U(t_j) \leq K_j U(0).$$

Now in this case there are at most $M-1$ zeros on I_k, so these zeros have to miss at least one D_j, say D_{j_n}, which means that U_n is nonnegaive and harmonic on D_{j_n}. Hence the preceding inequality can be applied with $U = U_n$ and $j = j_n$, by which we get from the assumption (7.3) that

$$U_n(0) \geq \frac{1}{K_{j_n}} U_n(t_{j_n}) \geq \frac{1}{K_{j_n}} n \log \beta \geq n \left(\min_{j=1,\ldots,M} \frac{1}{K_j} \right) \log \beta.$$

This gives us for large n

$$U_n(0) \geq M,$$

which, together with (7.7), proves our lemma because M can be arbitrary large.

■

Having this lemma at our disposal, we are proceeding with the construction of v_{k+1}.

Let us choose and fix an n_k such that $\theta_{n_k} < 1/k$ (see the previous lemma), and let $\mathcal{P}_k$ denote the set of polynomials P_n of degree at most n_k that satisfy

$$(7.8) \qquad w_k^{n_k}(x)|P_n(x)| \leq 1, \qquad x \in S_k.$$

This set of polynomials of a fixed degree is compact in the supremum norm, hence we get from Lemma 7.1 that there are finitely many points $t_1, \ldots, t_N$ on I_k such that if for $P_n \in \mathcal{P}_k$ the inequalities

$$(7.9) \qquad w_k^{n_k}(t_j)|P_n(t_j)|\beta^n \leq 1,$$

are satisfied for every j, then

$$(7.10) \qquad w_k^{n_k}(0)|P_n(0)| \leq \frac{1}{k}.$$

Let H_k be a subset of I_k containing the set $\{t_j\}_{j=1}^N$ such that it is symmetric with respect to the origin, it consists of finitely many intervals, and $\text{cap}(H_k)$ is some c_k that will we selected below together with the positive number ρ_k. If $\omega_k := \omega_{H_k}$ is the equilibrium measure on H_k, then the potential of the measure $\rho_k \omega_k$ is nonnegative on $[-1, 1]$ (by the symmetry of H_k), it is of the order $O(\rho_k)$ on S_k, at each t_j it has the value $\rho_k \log(1/c_k)$, and at every other point of $[-1, 1]$ its value is at most $\rho_k \log(1/c_k)$. Thus, by appropriately choosing c_k and ρ_k (to have, in particular $\rho_k \log(1/c_k) = \log \beta$), we can achieve that the weight $w_k^* = w_k \exp(U^{\rho_k \omega_k})$ will satisfy the property that

$$(7.11) \qquad (w_k^*)^{n_k}(x)|P_{n_k}(x)| \leq 1, \qquad x \in S_k \cup I_k,$$

implies

$$(7.12) \qquad (w_k^*)^{n_k}(0)|P_{n_k}(0)| \leq \frac{2}{k}$$

(see (7.8)–(7.10)), furthermore $|\log(w_k^*) - \log(w_k)| \leq \log\beta$. The point is that by adding the measure $\rho_k\omega_k$ to v_k we get a measure the potential of which is approximately the same as that of v_k on S_k (this is achieved by chosing ρ_k sufficiently small), but at the points t_j it is larger than the latter one by the amount $\log\beta$, and otherwise on $[-1,1]$ these two potentials differ by at most $\log\beta$. Furthermore, the argument also shows that there is a $\rho_k^* < 2^{-k}$ such that for every $0 < \rho_k < \rho_k^*$ there is an appropriate c_k (say $\rho_k \log c_k = \log\beta$) with the property that for the corresponding w_k^* (7.11) implies (7.12).

Now there is an $\epsilon_k > 0$ such that if w is any function satisfying

$$|\log(w(x)) - \log(w_k^*(x))| \leq \epsilon_k$$

for every $x \in \{0\} \cup S_k \cup H_k$, which is a subset of $\{0\} \cup S_{k+1}$, then

$$(7.13) \qquad w^{n_k}(x)|P_{n_k}(x)| \leq 1, \qquad x \in S_k \cup I_k,$$

implies

$$(7.14) \qquad w^{n_k}(0)|P_{n_k}(0)| \leq \frac{3}{k}.$$

Let h_k be the density of ω_k. The function $v_k + \rho_k h_k$ is almost the desired v_{k+1}, except that it has the following deficiencies: its integral is $1 + \rho_k$, that is slightly larger than 1, it has infinite singularities at the endpoints of the subintervals of H_k, and it is not everywhere positive on S_{k+1}. But all these are easy to rectify. In fact, it easily follows from the monoton convergence theorem and Dini's theorem about the uniform convergence of an increasing sequence of continuous functions provided the limit is continuous, that if we take $h_k^* := \min(h_k, M)$, then for large M the potentials of this new function and that of ω_k differ as little as we wish. Then we can distribute the mass $\int(h_k - h_k^*)$ that we saved in replacing h_k by h_k^* to $S_{k+1} \setminus (S_k \cup H_k)$ so as to get a continuous continuation of v_k that is smaller than $1/(k+1)$ if $|x| \leq 2^{-k}$ and takes the value $1/(k+3)$ at the points $\pm 2^{-k-1}$. In other words, there exists a continuous function $\overline{v}_{k+1}$ with the following properties: it coincides with v_k on S_k, coincides with h_k^* on H_k, continuous and positive on S_{k+1}, zero outside S_{k+1} (recall that $S_k \cup H_k \subset S_{k+1}$), is smaller than $1/(k+1)$ if $|x| \leq 2^{-k}$, takes the value $1/(k+3)$ at the points $\pm 2^{-k-1}$, and it has integral $1 + \rho_k$. Now

$$v_{k+1} := \frac{1}{1 + \rho_k}\overline{v}_{k+1}$$

with some small $\rho_k^* > \rho_k > 0$ clearly satisfies all the properties that we required of v_l for $l = k + 1$. Furthermore, this construction can be done in such a way that for the corresponding $w_{k+1} = \exp(U^{v_{k+1}})$ the inequality

$$(7.15) \qquad |\log(w_{k+1}(x)) - \log(w_k^*(x))| \leq \frac{\epsilon_k}{2}, \qquad x \in [-1,1],$$

is satisfied with the ϵ_k chosen before, furthermore

$$(7.16) \qquad |\log(w_{k+1}(x)) - \log(w_k(x))| \le 2\log\beta, \qquad x \in [-1,1].$$

Until now we have not said anything of the constant $\beta = \beta_k > 1$. To ensure the uniform convergence of the sequence $\{w_k\}$ on $[-1,1]$ let us require that $\log\beta_k = \min(\epsilon_{k-1}/8, 2^{-k})$, which, together with (7.16) yields the uniform convergence of $\{w_k\}$. We can clearly assume that $\epsilon_k \le \epsilon_{k-1}/4$ and $\epsilon_k \le 2^{-k}$ are also true for every k. Then

$$w(x) = \lim_{k\to\infty} w_k(x)$$

exists and continuous on $[-1,1]$ (actually on the whole real line), and it follows from (7.15), (7.16) and the inequality $\log\beta_l \le \epsilon_{l-1}/8$ that

$$(7.17) \qquad |\log(w(x)) - \log(w_k^*(x))| \le |\log(w_{k+1}(x)) - \log(w_k^*(x))|$$

$$+ \sum_{l>k} |\log(w_{l+1}(x)) - \log(w_l(x))| \le \frac{\epsilon_k}{2} + \frac{1}{4}\sum_{l>k}\epsilon_{l-1} < \epsilon_k$$

for every $x \in [-1,1]$.

Our construction gives that the sequence v_k uniformly converges on every compact subset of $(-1,1)$, and if

$$v(x) = \lim_{k\to\infty} v_k(x),$$

then v is positive on $(-1,0)\cup(0,1)$, continuous on $[-1,1]$, $v(0) = 0$, furthermore

$$w(x) = \lim_{k\to\infty} w_k(x) = \lim_{k\to\infty} \exp(U^{v_k}(x)) = \exp(U^v(x))$$

for at least every $x \in (-1,1)$. Thus, by Lemma 5.1 we get that the equilibrium measure associated with w will be $v(x)dx$, i.e. v is the density of μ_w.

Thus, it has left to prove that if a function f is uniformly approximable on $[-1,1]$ by weighted polynomials of the form $w^n P_n$, then f must vanish at the origin. In fact, without loss of generality assume that $|f| \le 1/2$. Let $\eta < 1/2$. If approximation is possible, then for large enough k there will be polynomials of degree at most n_k such that $|f - w^{n_k}P_{n_k}| \le \eta$ on $[-1,1]$. But then (7.13) is true in view of (7.17), hence (7.14) holds. Thus, we can conclude

$$|f(0)| \le |w^{n_k}(0)P_n(0)| + |f(0) - w^{n_k}(0)P_{n_k}(0)| \le \frac{3}{k} + \eta,$$

and since here $\eta > 0$ is arbitrary and k can be any large number, it follows that $f(0) = 0$ as we have claimed.

∎

In the previous construction we have considered the approximation problem on $[-1,1]$. To get an example when w is defined on $\mathbf{R}$ all we have to do is to extend w to $\mathbf{R} \setminus [-1,1]$ so that it be be admissible in the sense of (1.3) and be smaller than $\exp(U^v)$ there.

7.2   Example 4.6

In example 4.6 we have to construct a weight w on $[-1, 1]$ such that the support of the corresponding extremal measure μ_w is $[-1, 1]$, μ_w has continuous density in $(-1, 1)$ which vanishes at the origin, and still every continuous function f that is zero at ± 1 can be uniformly approximated by weighted polynomials of the form $w^n P_n$.

Here, as opposed to Example 4.5, we will construct a v which tends to 0 at the origin very slowly.

Let $\{f_j\}$ be a countable system of continuous functions that vanish at ± 1 such that the linear combinations of $\{f_j^2\}$ is dense in $C_0(-1, 1)$. Then it is enough to approximate each f_j^2. Without loss of generality we may assume that each f_j is nonnegative smaller than one at every point of $[-1, 1]$.

Let

$$\omega(x) = \frac{1}{\pi} \frac{1}{\sqrt{1 - x^2}}$$

be the so called arcsine distribution, which is nothing else than the equilibrium distribution of the interval $[-1, 1]$. For each n let ω_n be a function that is continous and nonnegative on $(-1, 1)$, coincides with ω on $(-1, -1/n) \cup (1/n, 1)$, vanishes at 0, it is at most 2 on $[-1/n, 1/n]$, and has integral 1. Choose $1 < \alpha < 2$, and let

$$\gamma = \left(\sum_{l=1}^{\infty} l^{-\alpha} \right)^{-1}.$$

With some sequence $\{n_k\}$ to be chosen below we set

$$(7.18) \qquad v(x) = \gamma \sum_{l=1}^{\infty} l^{-\alpha} \omega_{n_l}.$$

Then v will be a continuous function on $(-1, 1)$ that is positive everywhere except at the origin, where it vanishes. We will show that by appropriate choice of $\{n_k\}$ the weight

$$(7.19) \qquad w(x) := \exp(U^v(x))$$

satisfies all the requirements. Since Lemma 5.1 easily implies that the density of μ_w is exactly v, all we have to show is that every continuous function that vanishes at ± 1 can be uniformly approximated by weighted polynomials $w^n P_n$.

We shall determine the sequence $\{n_k\}$ recursively along with two other sequences $\{N_k\}$ and $\{M_k\}$ tending to infinity.

Suppose that $n_1, \ldots, n_{k-1}$, $N_1, \ldots, N_{k-1}$ and $M_1, \ldots, M_{k-1}$ have already been chosen, and for $m \geq 2$ let

$$v_{k,m}(x) = \gamma_k \left(\sum_{l=1}^{k-1} l^{-\alpha} \omega_{n_l} + k^{-\alpha} \omega_m + \beta_k \omega \right)$$

where

$$\beta_k = \sum_{l=k+2}^{\infty} l^{-\alpha} \quad \text{and} \quad \gamma_k = \left(\sum_{l \neq k+1} l^{-\alpha} \right)^{-1},$$

and

$$v_{k,\infty}(x) = \gamma_k \left(\sum_{l=1}^{k-1} l^{-\alpha} \omega_{n_l} + k^{-\alpha} \omega + \beta_k \omega \right).$$

Then, if we set $w_{k,m} = \exp(U^{v_{k,m}})$, it follows from Lemma 5.1 that $\mu_{w_{k,m}}$ has density $v_{k,m}$ (note that $v_{k,m}$ has integral 1).

The number m stands for the next n_l, i.e. for n_k. We shall choose m only at the very end of our construction. Our aim will be to get estimates that do not depend on the actual choice of m.

Let us follow the considerations of Sections 2 and 6. By the method of Section 2 the potential $nU_{k,m}$ of $nv_{k,m}$ can be approximated by potentials corresponding to the sum of n Dirac measures. We have to watch however the size of the remainder terms and have to carefully chose L because we want to get an estimate which is *uniform in* m. The details are as follows.

Let $\epsilon > 0$, and consider the set $J_\epsilon = [-1+\epsilon, 1-\epsilon]$ from Section 2, and the intervals I_j discussed there for the function $v_{k,m}$. The best constants c, C for which $c/n \leq |I_j| \leq C/n$ for every $I_j \cap J_{\epsilon^2} \neq \emptyset$ can be easily seen to satisfy $c \geq \epsilon$ and $C \leq 4\pi k^{\alpha-1}$, which follow from

$$\frac{1}{4\pi} k^{1-\alpha} \leq v_{k,m} \leq \frac{2}{\pi} \frac{1}{\sqrt{1-x^2}}.$$

Note that these estimates do not depend on m. Thus, if $L = k^{3\alpha/2-1}$, then in the present case we get for the sum in (2.8) the upper bound

$$(7.20) \quad C_1 \sum_j \frac{(C/n)^2}{(L/n)^2 + (c(j-j_0)/n)^2}$$

$$\leq C_1 \sum_{j=0}^{\infty} \frac{k^{2\alpha-2}}{k^{3\alpha-2} + \epsilon^2 j^2} + C_1 \max_j |I_j| \sum_{I_j \cap J_{\epsilon^2} = \emptyset} |I_j| \epsilon^{-2} \leq \frac{C_2}{\epsilon} k^{\alpha/2-1}$$

independently of m, for every sufficiently large n.

Let μ_n denote the measure that we obtain by translating $v_{k,m}(x)dx$ by iL/n (see Section 2). Recalling that $v_{k,m}$ and $v_{k,\infty}$ differ by at most $2k^{-\alpha}$, the argument of (2.5)–(2.6) yields that the potential of $v_{k,m}$ and of μ_n differ by

$$\frac{\pi L v_{k,\infty}(x)}{n} + \frac{L \chi_{k,m,n}(x)}{n} + o\left(\frac{L}{n}\right),$$

uniformly in m, where $|\chi_{k,m,n}(x)| \leq 2k^{-\alpha}$ (i.e. the $o(1)$ is uniform in m, and the estimate on $\chi_{k,m,n}$ is also independent of m). Note that because of $L = k^{3/2\alpha-1}$, in absolute value the second error term on the right is at most

$$\frac{L|\chi_{k,m,n}(x)|}{n} \leq \frac{2k^{3/2\alpha-1}k^{-\alpha}}{n} = \frac{2k^{\alpha/2-1}}{n},$$

which, after multiplication by n, is of the same order as the error term we obtained previously in (7.20) as the the analogue of the estimate for (2.8).

Thus, altogether we get for large n the following analogue of (2.1):

$$(7.21) \quad w_{k,m}^n(x)|Q_n(x)| = \exp(\pi k^{3\alpha/2-1}v_{k,\infty}(x) + \chi_{k,m,n}^*(x) + R_{k,m,n}(x))$$

for $x \in J_\epsilon$ and large n, where

$$|\chi_{k,m,n}^*(x)| := L|\chi_{k,m,n}(x)| \le 2k^{\alpha/2-1}$$

and

$$|R_{k,m,n}(x)| \le \frac{C_2}{\epsilon} k^{\alpha/2-1},$$

and these estimates are uniform in m. Hence, if we set $\epsilon = k^{(\alpha/2-1)/4}$, then it follows that
$$(7.22) \qquad |\chi_{k,m,n}^*(x)| \le 2k^{(\alpha/2-1)/4}$$
and
$$(7.23) \qquad |R_{k,m,n}(x)| \le C_2 k^{(\alpha/2-1)/4}.$$

This finishes the discussion of the approximation on J_ϵ.

Outside J_ϵ we used in Section 2 the estimate (2.2). Here the form of (2.2) we need is given by (6.10), i.e. we need it only in the form that

$$(7.24) \qquad w_{k,m}^n(x)|Q_n(x)| = e^{o(n)}$$

uniformly in m, which is clear from the uniform equicontinuity of the functions $w_{k,m}$ (see the proof of (6.10) in Section 6).

Now let us recall the definition of γ and γ_k from the beginning of the construction. Obviously, $\lambda_k := \gamma_k/\gamma > 1$. If we set

$$v_{k,m}^* = \gamma\left(\sum_{l=1}^{k-1} l^{-\alpha}\omega_{n_l} + k^{-\alpha}\omega_m + \beta_{k-1}\omega\right)$$

(note that now on the right we have β_{k-1} to have integral 1) and

$$w_{k,m}^* = \exp(U^{v_{k,m}^*}),$$

then it follows from the fact that the potential of ω is constant on $[-1,1]$, that $w_{k,m}$ and $(w_{k,m}^*)^{\lambda_k}$ differ only in a constant, hence together with (7.21)–(7.24) we get that for large n there are polynomials $Q_{[n/\lambda_k]}^*$ of degree at most $[n/\lambda_k]$ such that (c.f. also Sections 2 and 6)

$$(7.25) \qquad (w_{k,m}^*)^n(x)|Q_{[n/\lambda_k]}^*(x)| =$$

$$\exp(\pi k^{3\alpha/2-1}U^{v_{k,\infty}}(x) + (n - \lambda_k[n/\lambda_k])v_{k,m}^*(x) + \chi_{k,m,[n/\lambda_k]}^*(x) + R_{k,m,[n/\lambda_k]}(x))$$

for $x \in J_\epsilon$, and
$$(7.26) \qquad (w_{k,m}^*)^n(x)|Q_{[n/\lambda_k]}^*(x)| = e^{o(n)},$$

and these are uniform in m.

Let $1 \leq j \leq k$ arbitrary, and consider the function f_j from the system $\{f_j\}$ from the beginning of the proof. Using (7.25) and (7.26) instead of (6.9) and (6.10), we obtain with the method of Section 6 that we can multiply this $Q^*_{[n/\lambda_k]}$ by a suitable polynomial of degree at most $n - [n/\lambda_k]$ so that for the so obtained polynomial $P_{k,j,m,n}$ of degree at most n we have

$$|(w^*_{k,m})^n(x)|P_{k,j,m,n}(x)| \quad - \quad f_j(x)\exp(\chi^*_{k,m,[n/\lambda_k]}(x) + R_{k,m,[n/\lambda_k]}(x))|$$
$$\leq \quad C_3\eta_{j,k} + o(1), \qquad x \in [-1,1]$$

uniformly in m, where $\eta_{j,k}$ is the maximum of f_j on $[-1,1] \setminus J_{2\epsilon}$ (recall that ϵ was chosen to be equal to $k^{(\alpha/2-1)/4)}$, and C_3 is some fixed constant (in the preceding formula we assumed χ^* and $R_{k,m,n}$ extended outside J_ϵ so as (7.22) and (7.23) remain valid for all $x \in [-1,1]$). In fact, to be able to do this all we need is that the set of functions

$$\{\exp(sU^{v^*_{k,m}}(x)) \big| 0 \leq s \leq 1, \ m = 1, 2, \ldots\}$$

is compact (see also the corresponding arguments in Sections 2 and 6), which is obvious from the definition of the functions ω_m. If we also take into account (7.22) and (7.23) and that the f_j's are at most 1 in absolute value, we can finally conclude that

$$|(w^*_{k,m})^n|P_{k,j,m,n}| - f_j| \leq C_4(\eta_{j,k} + k^{(\alpha/2-1)/4)}) + o(1)$$

with an absolute constant C_4, where $o(1) \to 0$ uniformly in m as $n \to \infty$.

Thus, there exists an $N_k > N_{k-1}$ such that if $n \geq N_k$, then independently of the choice of m we have

$$|(w^*_{k,m})^n|P_{k,j,m,n}| - f_j| \leq 2C_4(\eta_{j,k} + k^{(\alpha/2-1)/4})$$

for $1 \leq j \leq k$, and this is how we choose N_k.

Now let us return to

$$v^*_{k,m} = \gamma\left(\sum_{l=1}^{k-1} l^{-\alpha}\omega_{n_l} + k^{-\alpha}\omega_m + \sum_{l=k+1}^{\infty} l^{-\alpha}\omega\right).$$

As we have already mentioned, here m stands for n_k, so the only difference between $v^*_{k,m}$ and v (see (7.18)) is that in the latter one the terms $l^{-\alpha}\omega$ with $l \geq k$ are replaced by $l^{-\alpha}\omega_{n_l}$. Consider now only those n for which $N_{k-1} \leq n < N_k$. If we replace the terms $l^{-\alpha}\omega$, $l > k$ by some $l^{-\alpha}\omega_{n_l}$ in $v^*_{k,m}$, then if these n_l's are sufficiently large, we get from the preceding estimate applied to $k-1$ rather than to k (in which case $v^*_{k-1,n_{k-1}}(x) = v^*_{k,\infty}(x)$) that

$$(7.27) \qquad |w^n|P_{k-1,j,n_{k-1},n}| - f_j| \leq 3C_4(\eta_{j,k-1} + (k-1)^{(\alpha/2-1)/4})$$

will be true for every $N_{k-1} \leq n < N_k$ (note that N_k has already been fixed) and $1 \leq j < k$. Thus, there exists an $M_k > M_{k-1}$ such that if $m = n_k, n_{k+1}, n_{k+2}, \ldots$

are all larger than M_k, then for w defined by (7.19) the estimate (7.27) is true. This gives the choice M_k.

Now all we have to do is to select $n_k = m > M_k$ and $n_k > n_{k-1}$. We emphasize again that this choice does not influence the choice of N_k and M_k, so our parameters can be selected in the order

$$\ldots, n_{k-1}, N_k, M_k, n_k, N_{k+1}, M_{k+1}, n_{k+1}, \ldots.$$

We claim that this choice is appropriate. Let $j \geq 1$ be arbitrary, and $n > j$ some large number. Then there is a k such that $N_{k-1} \leq n < N_k$. (7.27) implies that

$$|w^n|P_{k-1,j,n_{k-1},n}| - f_j| \leq 3C_4(\eta_{j,k-1} + (k-1)^{(\alpha/2-1)/4}).$$

Here k tends to infinity as n does so, and then the right hand side tends to zero, hence the uniform approximability of f_j^2 by the weighted polynomials $w^{2n}|P_{k-1,j,n_{k-1},n}|^2$ follows. Since the linear combinations of the f_j^2's is dense in $C_0(-1,1)$, it follows that uniform approximation (on $[-1,1]$) of every continuous f with $f(\pm 1) = 0$ is possible by weighted polynomials of the form $w^{2n}P_{2n}$. But this implies the uniform approximability by weighted polynomials of odd degree exacly as we mentioned it in Section 2 (divide through by w, approximate and multiply back).

Part III

Varying weights

In several problems weighted polynomials of the form $W_n P_n$ appear where $\{W_n\}$ is a sequence of weights (see e.g. Section 12), i.e. the weights are not powers of a fixed weight function. In such a case we set $w_n = W_n^{1/n}$ and consider weighted polynomials $w_n^n P_n$ with varying weights w_n. The method of the preceding sections allows us to deduce convergence results in this setting. The applications of these results will be given in Chapter IV.

8 Uniform approximation by weighted polynomials with varying weights

Let us begin with the analogue of Theorem 4.2.

Theorem 8.1 *Suppose that $\{w_n\}$ is a sequence of weights such that the extremal support S_{w_n} is $[0,1]$ for all n, and let O be an open subset of $(0,1)$ for which the set $[0,1]\backslash O$ is of zero capacity. If the equilibrium measures μ_{w_n} are absolutely continuous with respect to Lebesgue measure on O: $\mu_{w_n}(x) = v_n(x)dx$, and the densities v_n are uniformly equicontinuous and uniformly bounded from below by a positive constant on every compact subset of O, then every continuous function that vanishes outside O can be uniformly approximated on $[0,1]$ by weighted polynomials $w_n^n P_n$, $\deg P_n \leq n$.*

Actually, the sequence $\{w_n^n P_n\}$ can be constructed in such a way that the convergence $w_n^n P_n \to f$ holds uniformly on some larger set $[-\theta, 1+\theta]$, $\theta > 0$ (provided of course the weights are defined there). This easily follows from the proof.

The conclusion is false for every O for which $[-1,1] \setminus O$ is not of positive capacity, but we shall not show this. For conditions in terms directly on w_n themselves which guarantee the assumptions in the theorem see Theorem 8.3 after the proof.

Proof of Theorem 8.1. We set $\Sigma = J^* = [0,1]$, $J_\epsilon = \{x \mid (x - \epsilon, x + \epsilon) \subseteq O\}$, and copy the proof of Theorem 4.2. This can be done word for word with one exception, and this is the estimate (6.10). In fact, the heart of the proof of Theorem 4.2 is Lemma 5.8 which is valid in the following form with virtually the same proof: *If v_n are uniformly equicontinuous on an interval $[x_0 - \epsilon_1, x_0 + \epsilon_1]$ and $v_n(t) \geq \epsilon_0$ there, then for $\lambda \leq 1/(1 - \epsilon_0\epsilon_1/2)$ the interval $[x_0 - \epsilon_1/2, x_0 + \epsilon_1/2]$ belongs to the interior of $S_{w_n^\lambda}$, and the densities $v_{w_n^\lambda}$ are also uniformly equicontinuous there.*

What goes wrong with (6.10)? In the present case we could claim (6.10) only under the assumption that the potentials $U^{\mu_{w_n}}(z)$ are uniformly equicontinuous on $[0, 1]$ as functions of the complex variable z (c.f. (2.7), which has been replaced in the proof of Theorem 4.2 by a similar relation where the left hand side is $o(1)$), and this may not be true.

In any case we have (6.10) in the form

$$(8.1) \qquad\qquad w_n^n(x)|Q_n(x)| \leq C_0 e^{C_0 n}.$$

Note however, that the densities are uniformly bounded on compact subsets of O, which easily implies that (6.10) is true inside O, i.e.

$$(8.2) \qquad\qquad w_n^n(x)|Q_n(x)| = e^{o(n)}$$

uniformly on compact subsets of O. In Section 6 the estimate (6.10) is used in conjunction with the estimates (6.13)–(6.15), i.e. with

$$(8.3) \qquad\qquad |R_m(x) - 1| \leq c^m, \qquad \text{for } x \in J_{2\epsilon},$$

$$(8.4) \qquad\qquad |R_m(x)| \leq c^m, \qquad \text{for } x \in J^* \setminus J_\epsilon,$$

and

$$(8.5) \qquad\qquad 0 \leq R_m(x) \leq 1, \qquad \text{for } x \in J_\epsilon \setminus J_{2\epsilon},$$

where c was *some* positive number less than 1. Now if this c was actually smaller than $\exp(-C_0)$ from (8.1), then the proof in Section 6 would be valid in the present case, as well. The proof also shows that any J_η with some fixed but small $\eta > 2\epsilon$ can stand in (8.3) and (8.5) instead of $J_{2\epsilon}$, furthermore in (8.3) we do not actually need geometric convergence, i.e. (8.3) can be replaced by

$$(8.6) \qquad\qquad |R_m(x) - 1| = o(1) \qquad \text{for } x \in J_\eta,$$

as $m \to \infty$. It is also clear from the proof that (8.5) can be replaced by the uniform boundedness of $R_m(x)$:

$$(8.7) \qquad\qquad |R_m(x)| \leq C \qquad \text{for all } m \text{ and } x \in J^*.$$

Thus, it is enough to show that in the present case for *arbitrary $\eta > 0$ and $c > 0$* we can choose an $\epsilon > 0$, such that (8.4), (8.6) and (8.7) hold for some polynomials R_m of degree at most m whenever m is sufficiently large.

The assumption that $[0, 1] \setminus O$ has zero capacity implies that the capacity of $J^* \setminus J_\epsilon$ tends to zero, hence our claim follows from the next lemma by setting $S = J_\eta$ and $K = J^* \setminus J_{\eta/2}$ if we apply it to the sets $L = J^* \setminus J_\epsilon$ with $\epsilon < \eta/2$, $\epsilon \to 0$. Thus, the verification of the lemma will complete the proof of Theorem 8.1.

Lemma 8.2 *Let S and K be two disjoint compact subsets of $[0, 1]$. Then there is a constant $\delta > 0$ such that for all compact subsets L of K and sufficiently large n there are polynomials P_n of degree at most n such that*

$$(8.8) \qquad\qquad |P_n(x)| \leq 2, \qquad x \in [0, 1],$$

$$(8.9) \qquad\qquad |P_n(x) - 1| \leq \left(\frac{1}{2}\right)^{\delta n}, \qquad x \in S$$

and

$$(8.10) \qquad\qquad |P_n(x)| \leq (\text{cap}(L))^{\delta n}, \qquad x \in L.$$

Proof. Let

$$T_m(z) = \prod_{j=1}^{m} (z - z_j)$$

be the polynomial of degree m that has all its zeros z_j in L and minimizes the norm $\|T_m\|_L$ among all such polynomials (these are the so called restricted Chebyshev polynomials associated with the set L). The classical proofs for the Chebyshev polynomials given e.g. in [51, Theorem III. 26] can be easily modified so that we obtain

(8.11) $$\lim_{m \to \infty} \|T_m\|_L^{1/m} = \mathrm{cap}(L).$$

Since all the zeros of T_m belong to $[0,1]$, we also have

(8.12) $$|T_m(x)| \leq 1, \quad x \in [0,1].$$

Let now S_ρ and K_ρ be the set of points on the plane the distance of which to S and K, respectively is at most ρ, and choose ρ so small that the sets S_ρ and K_ρ be disjoint. Consider the function $f_m(z)$ which is defined to be 1 on K_ρ and $1/T_m(z)$ on S_ρ. This f_m is analytic on $S_\rho \cup K_\rho$ and we have the bound

$$|f_m(z)| \leq \left(\frac{1}{\mathrm{dist}(S,K)} \right)^m =: C_1^m$$

for it, hence by a classical approximation theorem of Bernstein ([52, Theorem 5, Sec. 4.5, p. 75]) there is a $\tau < 1$ and there are polynomials R_k of degree at most k such that

$$|R_k(z) - f_m(z)| \leq C_1^m \tau^k, \quad z \in S_{\rho/2} \cup K_{\rho/2}.$$

We set here $k = rm$, where r is so large that $\tau^r < 1/4C_1$ holds. Thus,

(8.13) $$|R_k(z) - f_m(z)| \leq 4^{-m}, \quad z \in S_{\rho/2} \cup K_{\rho/2}.$$

We also get from the Bernstein–Walsh lemma ([52, p. 77]) that with some constant $C_2 \geq 1$

(8.14) $$|R_k(x)| \leq C_2^k C_1^m =: C_3^m, \quad x \in [0,1]$$

(note that here $C_3 \geq 1$).

We have already used in (6.13)–(6.15) the consequence of [15, Theorem 3] that if there are two disjoint systems of subintervals of $[0,1]$, then there are polynomials that take values in between 0 and 1 on $[0,1]$ and geometrically converges to zero respectively to 1 on the two systems of intervals. Thus, we can choose a κ and for all sufficiently large l polynomials Q_l of degree at most l such that

$$|Q_l(x)| \leq \kappa^l, \quad x \in [0,1] \setminus S_{\rho/2},$$

and

$$|Q_l(x) - 1| < \kappa^l, \quad x \in S.$$

We set here $l = sm$ with an s such that $\kappa^s < 1/2C_3$ holds, by which we get

$$(8.15) \qquad |Q_l(x)| \leq \left(\frac{1}{2C_3}\right)^m, \qquad x \in [0,1] \setminus S_{\rho/2}$$

and

$$(8.16) \qquad |Q_l(x) - 1| < \left(\frac{1}{2}\right)^m, \qquad x \in S.$$

Finally, we set $P_n(x) = T_m(x)R_k(x)Q_l(x)$ which has degree at most $(r+s+1)m$.

On S we have $f_m(x)T_m(x) = 1$, hence

$$|R_k(x)T_m(x) - 1| = |(R_k(x) - f_m(x))T_m(x)| \leq 4^{-m} \cdot 1 \leq 2^{-m}$$

by (8.13) and (8.12). If we take into account (8.16) then we can conclude that $|P_n(x) - 1| \leq 3 \cdot 2^{-m}$ on S, which proves (8.9).

In the same fashion, on L the product $|R_k(x)Q_l(x)|$ is at most 2^{-m} by (8.14) and (8.15), and we have $|T_m(x)| < (\text{cap}(L))^{m/2}$ for large enough m, hence $\|P_n\|_L \leq (\text{cap}(L))^{m/2}$ proving (8.10).

If $x \notin S_{\rho/2}$, then (8.12), (8.15) and (8.14) imply that $|P_n(x)| \leq 1 < 2$. Finally, if $x \in S_{\rho/2}$ then the same conclusion follows from (8.16), (8.12) and (8.13), because the latter two imply

$$|T_m(x)R_k(x) - 1| = |T_m(x)||R_k(x) - f_m(x)| \leq 4^{-m}.$$

These prove (8.8), and the proof is complete.   ∎

It is clear how we should modify the proof in order to achieve convergence on some $[-\theta, 1+\theta]$: all we need to do is to add the sets $[-\theta, 0]$ and $[1, 1+\theta]$ to $L = J^* \setminus J_\epsilon$. As $\epsilon, \theta \to 0$ the capacity of the new L will tend to zero, and this is all we needed above.   ∎

Now let us discuss some conditions that guarantee directly in terms on the weights w_n that Theorem 8.1 can be applied. We shall always assume that the weights are normalized so that the support S_{w_n} of the corresponding equilibrium measure is $[0,1]$.

Theorem 8.3 *Suppose that $\{w_n\}$, $w_n = \exp(-Q_n)$ is a sequence of weights such that the extremal support S_{w_n} is $[0,1]$ for all n, on every closed subinterval $[a,b] \subset (0,1)$ the functions $\{Q_n\}$ are uniformly of class $C^{1+\epsilon}$ for some $\epsilon > 0$ that may depend on $[a,b]$, and the functions $tQ'_n(t)$ are nondecreasing on $(0,1)$ and there are points $0 < c < d < 1$ and an $\eta > 0$ such that $dQ'_n(d) \geq cQ'_n(c) + \eta$ for all n. Then every continuous function that vanishes outside $(0,1)$ can be uniformly approximated on $[0,1]$ by weighted polynomials $w_n^n P_n$, $\deg P_n \leq n$.*

Being uniformly in $C^{1+\epsilon}$ means that the derivatives satisfy uniformly a Lipshitz condition

$$|Q_n'(x) - Q_n'(y)| \leq C|x - y|^\epsilon, \qquad x \in [a, b], \ y \in (0, 1)$$

with constants C and $\epsilon = \epsilon_{a,b} > 0$ independent of x and y. Note that our assumptions require only $C^{1+\epsilon}$ smoothness on Q_n inside $(0, 1)$.

We can conclude again that $w_n^n P_n \to f$ holds uniformly on some larger set $[-\theta, 1 + \theta]$, $\theta > 0$ (provided of course the weights are defined there).

Corollary 8.4 *Suppose that $\{w_n\}$, $w_n = \exp(-Q_n)$ is a sequence of even weights such that the extremal support S_{w_n} is $[-1, 1]$ for all n, and on $[0, 1]$ the functions satisfy the conditions of the preceding theorem. Then every continuous function that vanishes outside $(-1, 1)$ and also at the point 0 is the uniform limit on $[-1, 1]$ of weighted polynomials $w_n^n P_n$, $\deg P_n \leq n$.*

What happens around 0 (i.e. what is the situation if the function to be approximated does not vanish at the origin) is quite complicated (see Theorems 12.2 and 12.3 for more details): if $w_n(x)$ for all n is the Freud weight $\exp(-\gamma_\alpha^{1/\alpha}|x|^\alpha)$, $\alpha > 0$, then clearly all the conditions of the Corollary are satisfied, but an(y) f that vanishes outside $(-1, 1)$ but not at the origin is approximable by weighted polynomials $w_n^n P_n$ if and only if $\alpha \geq 1$ (see [30], and also the discussion in Section 12).

We begin the proof of Theorem 8.3 by a lemma.

Lemma 8.5 *Let $w(x) = \exp(-Q(x))$ be such that $S_w = [0, 1]$ and that $tQ'(t)$ nondecreases on $(0, 1)$. Then the density of the equilibrium measure $d\mu_w(t) = v(t)dt$ is given by*

$$(8.17) \quad v(t) = \frac{1}{\pi^2}\sqrt{\frac{1-t}{t}}\int_0^1 \frac{sQ'(s) - tQ'(t)}{s - t}\frac{1}{\sqrt{s(1-s)}}ds + \frac{D}{\sqrt{t(1-t)}},$$

where

$$D = \frac{1}{\pi} - \frac{1}{\pi^2}\int_0^1 \sqrt{\frac{s}{1-s}}Q'(s)ds,$$

and here $D \geq 0$.

Proof. Let $f(x) = Q(x^2)/2$, $x \in [-1, 1]$. It was shown in [28, Lemma 5.1] that the integral equation

$$\int_{-1}^1 \log\frac{1}{|x - t|}g(t)dt = -f(x) + C$$

where C is some constant has a solution $g(t)$ of the form

$$g(t) = \frac{2}{\pi^2}\sqrt{1 - t^2}\int_0^1 \frac{sf'(s) - tf'(t)}{(1 - s^2)^{1/2}(s^2 - t^2)}ds + \frac{D_1}{\sqrt{1 - t^2}},$$

where

$$D_1 = \frac{1}{\pi} - \frac{1}{\pi^2} \int_{-1}^{1} \frac{sf'(s)}{\sqrt{1-s^2}},$$

furthermore g is even and has total integral 1 over $[-1,1]$. If we set $h(t) = g(\sqrt{t})/\sqrt{t}$, $t \in [0,1]$ then h will have total integral 1 over $[0,1]$, and by the symmetry of g its potential satisfies

$$\int_0^1 \log \frac{1}{|x-u|} h(u)du \;=\; 2\int_0^1 \log \frac{1}{|x-t^2|} g(t^2)dt = 2\int_{-1}^1 \log \frac{1}{|\sqrt{x}-t|} g(t)dt$$

$$= -2f(\sqrt{x}) + C = -Q(x) + C$$

for every $x \in [0,1]$. On applying Lemma 5.1 we can conclude that $d\mu_w(t) = h(t)dt$ i.e. $h(t) = v(t)$ *provided* we can show that h is nonnegative (to be more precise the equality $d\mu_w(t) = h(t)dt$ automatically follows from the properties above, but we shall need to prove the nonnegativity of D anyway, and this easily implies the nonnegativity of h). The nonnegativity of h will follow from the last statement of the lemma that we are going to show in a short while. In fact, we can see by integration by parts that the two constants D and D_1 are the same, so the relation $D \geq 0$ is the same as $D_1 \geq 0$. Furthermore, the first term in the expression of g is nonnegative in view of the fact that $sQ'(s)$ nondecreasing. Hence, the nonnegativity of D implies that of h, and this, as we have seen before, implies $v(t) = h(t)$. Now if we carry out the substitution $f(s) = Q(s^2)/2$ and $u = s^2$ in the formula for g, we obtain the form of v stated in the lemma.

Thus, it has left to show that $D \geq 0$, i.e.

$$(8.18) \qquad\qquad \frac{1}{\pi} \int_0^1 \sqrt{\frac{s}{1-s}} Q'(s)ds \leq 1.$$

First we show that for every $t \in (0,1)$

$$(8.19) \qquad\qquad \frac{1}{2\pi} \int_0^1 \log \left|\frac{1-t}{s-t}\right| \frac{1}{s^{1/2}(1-s)^{3/2}} ds = 1.$$

We shall do this by examining the integral

$$I_\epsilon = \frac{1}{2\pi} \int_0^{*1} \log \left|\frac{1-t}{s-t}\right| \frac{1}{s^{1/2}(1-s)^{3/2}} ds = 1,$$

where $*$ indicates that we skip a small $(t-\epsilon, t+\epsilon) \subseteq (0,1)$ neighborhood of t during integration. If

$$\chi_{\epsilon,t}(u) = \begin{cases} (u-t)^{-1} & \text{if } |t-u| \geq \epsilon \\ 0 & \text{otherwise,} \end{cases}$$

then

$$\log \left|\frac{1-t}{s-t}\right| = \int_s^1 \chi_{\epsilon,t}(u)du$$

for all s which belongs to the range of integration in I_ϵ, hence integration by parts yields

$$I_\epsilon = \frac{1}{\pi} \int_0^1 \sqrt{\frac{s}{1-s}} \chi_{\epsilon,t}(s) ds.$$

As $\epsilon \to 0$ this tends to the principal value integral

$$\frac{1}{\pi} \mathrm{PV} \int_0^1 \frac{s}{s-t} \frac{1}{\sqrt{s(1-s)}} ds.$$

Unsing now that $s/(s-t) = 1 + t/(s-t)$ and that

$$(8.20) \qquad \mathrm{PV} \int_0^1 \frac{1}{s-t} \frac{1}{\sqrt{s(1-s)}} ds = 0,$$

for all $t \in (0,1)$ (see [39, p. 251, (88.9)]) we obtain (8.19).

Since in (8.19) the integrand is negative to the left of $2t - 1$ and positive on $(2t - 1, 1)$, we obtain for all $a < 1$ and $t \in (0,1)$ the inequality

$$\frac{1}{2\pi} \int_0^a \log \left| \frac{1-t}{s-t} \right| \frac{1}{s^{1/2}(1-s)^{3/2}} ds \le 1.$$

Here the integrand is bounded from below by the function

$$C_a \frac{1}{\sqrt{s}} \min \left\{ 0, \log \frac{1-t}{t} \right\}, \qquad C_a = (1-a)^{-3/2},$$

integrable with respect to the product measure $ds\, d\mu_w(t)$, hence we can integrate this inequality with respect to $d\mu_w(t)$, and we can change the order of integration. Taking into account that the potential of μ_w equals $-Q + \mathrm{const}$ on $(0,1)$ (see Theorem A in Section 1), we obtain that

$$\frac{1}{2\pi} \int_0^a \frac{Q(1) - Q(s)}{s^{1/2}(1-s)^{3/2}} ds \le 1.$$

Our assumptions imply that Q is monotone in a left neighborhood of 1 ($tQ'(t)$ is either positive or negative in such a neighborhood), so by the monotone convergence theorem we can conclude from the preceding inequality the same for $a = 1$:

$$\frac{1}{2\pi} \int_0^1 \frac{Q(1) - Q(s)}{s^{1/2}(1-s)^{3/2}} ds \le 1,$$

and from here (8.18) follows by integration by parts (verify this separately for the cases when $tQ'(t)$ is negative or positive, respectively in a left neighborhood of 1).

After that we can turn to the

Proof of Theorem 8.3. All we have to do is to show that the conditions of Theorem 8.1 are satisfied with $O = (0,1)$.

Under the conditions of Theorem 8.3 the functions $sQ'_n(s)$ uniformly belong to Lipϵ for some ϵ on every compact subinterval of $[a, b] \subset (0,1)$, and it is well known from the theory of the singular integrals with Cauchy kernels (see e.g. the Plemelj–Privalov theorem in [39, p. 46]) that then the same is true of the integrals in (8.17) (with Q replaced by Q_n, of course). Hence, the uniform equicontinuity of the densities v_n on compact subsets of $(0,1)$ has been established.

It has remained to show that they are uniformly bounded away from zero on every compact subset of $(0,1)$. Since the second terms in (8.17) are nonnegative by the last statement of the lemma, it is enough to show that even the first terms in (8.17) (again with Q replaced by Q_n) remain above a positive constant on $(0,1)$.

We set $g(t) = tQ'_n(t)$, and it is enough to show that if g is nondecreasing on $(0,1)$ and $g(d) \geq g(c) + \eta$, then for all $t \in (0,1)$

$$\int_0^1 \frac{g(s) - g(t)}{s - t} \frac{1}{\sqrt{s(1-s)}} ds \geq \theta,$$

where $\theta > 0$ depends only on c, d and η. In fact, it is enough to prove this for continuously differentiable g, in which case the claim follows from the fact that

$$\int_0^1 \frac{g(s) - g(t)}{s - t} \frac{1}{\sqrt{s(1-s)}} ds \geq \int_0^1 \frac{g(s) - g(t)}{s - t} ds,$$

and here

$$
\begin{aligned}
\int_0^1 \frac{g(s) - g(t)}{s - t} ds &= \int_0^1 \frac{1}{s - t} \int_s^t g'(u) du\, ds \\
&= \int_0^t g'(u) \log \frac{t}{t - u} du + \int_t^1 g'(u) \log \frac{1 - t}{u - t} du \\
&\geq \alpha \int_c^d g'(u) du = \alpha(g(d) - g(c)) \geq \alpha \eta
\end{aligned}
$$

with

$$\alpha = \min \left\{ \log \frac{1}{1 - c}, \log \frac{1}{d} \right\}.$$

■

The Corollary immediately follows from Theorems 8.1 and 8.3 (applied to the interval $[-1, 1]$ rather than to $[0, 1]$ and to the set $O = (-1, 0) \cup (0, 1)$) if we use the substitution applied in the proof of Lemma 8.5.

9   Modification of the method

In this section, we modify the method that we used above. This modification allows us to get better aproximation around the endpoints of the interval. These are needed if we want to handle infinite singularities inside the support of the generating measure, or if the approximation has a second constraint that frequenly appears in applications (see Theorem 10.1 and 10.2 below).

The modification is roughly as follows: In the first two parts of this work we approximated a given potential by first translating the generating measure and then appropriately discretizing it. The idea here is similar, but instead of translation, that can be viewed as projection onto a segment, we shall project it onto a curve, which is closer to the support exactly where the generating measure is larger.

We start with a technical lemma.

Lemma 9.1 *Suppose that* $\{u_n\}$ *is a sequence of nonnegative functions on* $(0,1)$ *satisfying the following conditions:* $\{u_n\}$ *is uniformly equicontinuous on every compact subinterval of* $(0,1)$,

$$\int_0^1 u_n = 1,$$

and for some constants $A > 0$, $\beta > -1$ *and* β_0

$$(9.1) \qquad u_n(t) \le A(t(1-t))^\beta, \qquad t \in (0,1),$$

$$(9.2) \qquad u_n(t) \ge \frac{1}{A}(t(1-t))^{\beta_0}, \qquad t \in (0,1).$$

Then there is an $L_0 > 1$ *such that for every* $L > L_0$ *there are polynomials* Q_n *of degree at most* n *such that for large* n, *say* $n \ge n_L$

$$(9.3) \quad 0 \;\le\; \log|Q_n(x)| + nU^{u_n}(x)$$
$$\le \begin{cases} BL^2 \log n & \text{if } x \in [0, n^{-1}] \cup [1 - n^{-1}, 1] \\ BL^2 \log 1/(x(1-x)) & \text{if } n^{-1} \le x \le 1 - n^{-1}, \end{cases}$$

and with some continuous functions g_n *that are uniformly bounded and uniformly equicontinuous on compact subsets of* $(0,1)$

$$(9.4) \quad |\log|Q_n(x)| + nU^{u_n}(x) - Lg_n(x)| \le BL^{-1} \qquad \text{if } L^{-6} \le x \le L^{-6}.$$

Here B is an absolute constant depending only on A, β *and* β_0.

The same conclusion holds if we assume the inequality in (9.2) only on the interval $[n^{-\tau}, 1 - n^\tau]$.

The lemma is true in more general situations (allowing several intervals or zero or infinite singularities in the weight), we shall comment on that after the proof.

Proof. We shall only concentrate on the behavior on the interval $[0, 1/2]$, the other half being symmetric. We shall indicate at the very end of the proof the necessary modifications we have to make to cover the whole interval $[0, 1]$.

Without loss of generality we can assume that $\beta < 0$ and $\beta_0 > 0$.

Fix a constant L. We select a continuously differentiable function v_n that has the same size as u_n on $[L^{-9}, 1 - L^{-9}]$, as follows: for $x \in [L^{-10}, 1 - L^{-10}]$ we set

$$(9.5) \qquad\qquad v_n(x) = \frac{1}{2d_L} \int_{x-d_L}^{x+d_L} u_n(t) \, dt,$$

where the constant d_L satisfies

$$(9.6) \quad \frac{1}{2} \le u_n(t_1)/u_n(t_2) \le 2 \qquad \text{for } t_1, t_2 \in [L^{-11}, 1 - L^{-11}], \ |t_1 - t_2| \le d_L$$

for all n (such a d_L exists because the functions u_n are uniformly equicontinuous and uniformly bounded below by a positive constant on compact subsets of $(0, 1)$).

Let $0 < \tau_0 < 1/(\beta_0 + 1)$ if (9.2) is true for all $t \in (0, 1)$, and $0 < \tau_0 < \min\{\tau, 1/(\beta_0 + 1)\}$ if we assume (9.2) only for $t \in [n^{-\tau}, n^{\tau}]$ (in the former case we may set $\tau = \infty$).

For an n let us choose consecutive intervals I_j, $j = 0, 1, \ldots, n - 1$ starting at 0 such that on I_j the function u_n has integral equal to $1/n$:

$$\int_{I_j} u_n = \frac{1}{n}.$$

As in Section 2 let ξ_j be the weight point of u_n on I_j:

$$\xi_j = n \int_{I_j} t u_n(t) dt.$$

Let $j = 0, 1, \ldots, N$ be those indices for which $I_j \subseteq [0, L^{-9}]$. Set now

$$(9.7) \qquad\qquad m_j = \begin{cases} |I_j| & \text{if } j = 0, 1, \ldots, N \\ \int_{I_j} u_n(t)/v_n(t) \, dt & \text{if } j > N. \end{cases}$$

We claim that the polynomials

$$Q_n(x) = \prod_{j=0}^{n-1} (x - \xi_j + iLm_j)$$

satisfy all the requirements of the lemma.

Before we embark on the proof, we make a few preliminary observations. In doing so let us agree that in what follows c, C denote positive absolute constants depending exclusively on A, β and β_0. However, we allow C and c to change from line to line. As before, the symbol $R \sim S$ will indicate that $c \le R/S \le C$ with some such c, C.

It follows from assumption (9.1) that

$$(9.8) \quad \xi_j \geq c \left(\frac{j+1}{n} \right)^{1/(1+\beta)}, \qquad |I_j| \geq c(j+1)^{-\beta/(1+\beta)} n^{-1/(1+\beta)},$$

and for all j

$$(9.9) \qquad m_j \sim |I_j|, \qquad 0 \leq j < n.$$

(9.2) gives

$$(9.10) \qquad \max_j |I_j| \leq C \max\{n^{-\tau}, n^{-1/(\beta_0+1)}\} \leq Cn^{-\tau_0}$$

regardless if (9.2) holds for all t or only for $t \in [n^{-\tau}, 1 - n^{-\tau}]$.

The choice of τ_0 implies via (9.10) that if $x \in [n^{-\tau_0}, 1/2]$ and I is an interval of length $\geq x/8$ that intersects $(x/2, 2x)$, then

$$\int_I u_n > \frac{1}{n}.$$

This means that there is at least two-two I_j lying stricly in the intervals $(x/2, x)$ and $(x, 2x)$. For the length of every such an I_j we obtain from (9.1) and (9.2)

$$(9.11) \qquad \frac{1}{A}(4x)^{-\beta}/n \leq |I_j| \leq A(x/4)^{-\beta_0}/n, \qquad I_j \subseteq (x/2, 2x).$$

Finally, we mention the formula

$$(9.12) \quad \log |Q_n(x)| + U^{u_n}(x) = \sum_{j=0}^{n-1} n \int_{I_j} \log \left| \frac{x - \xi_j + iLm_j}{x - t} \right| u_n(t) \, dt.$$

The lower estimate

First we prove that

$$\log |Q_n(x)| + U^{u_n}(x) \geq 0.$$

In view of (9.12) it is enough to show that

$$n \int_{I_j} \log \left| \frac{x - \xi_j + iLm_j}{x - t} \right| u_n(t) \, dt \geq 0$$

for all x and j.

Let $x \in I_{j_0}$. If $j = j_0$, then the quantity under the logarithm is at least as large as

$$L \left| \frac{m_j}{x - t} \right| \geq L \frac{m_j}{|I_j|} \geq 1,$$

for sufficiently large L, where we used (9.9).

If, however, $j \neq j_0$, then the quantity under the logarithm is at least as large as

$$\left| \frac{x - \xi_j}{x - t} \right|,$$

and by the concavity of the logarithm function $\log|x - t|$ for $t \in I_j$ we get

$$n \int_{I_j} \log|x - t| u_n(t)\, dt \le \log|x - \xi_j|.$$

These inequalities prove the lower estimate.

The upper estimate

We distinguish two cases according to the location of x.

Case I. $x \in [0, n^{-r_0}]$. Let $x \in I_{j_0}$. We divide the sum in (9.12) into three ranges:

$$\sum_{j<j_0-1} + \sum_{j=j_0-1}^{j_0+1} + \sum_{j>j_0+1} = \sum_1 + \sum_2 + \sum_3.$$

We estimate these three sums separately.

In $\sum_2$ we have (at most) three terms. The second one is at most as large as

$$n \int_{I_{j_0}} \log \left| \frac{x - \xi_{j_0} + iLm_{j_0}}{x - t} \right| u_n(t)\, dt \le n \int_{I_{j_0}} \log \frac{||I_{j_0}| + iCL|I_{j_0}||}{|x - t|} u_n(t)\, dt$$

$$\le \frac{1}{2} \log((1 + C^2 L^2)|I_{j_0}|) + n \int_{I_{j_0}} \log \frac{1}{|x - t|} u_n(t)\, dt.$$

We break the last integral into two parts, for integration over $|x - t| \le n^{-(1+\beta)}$ and the rest. In view of (9.1) we can get the following upper bound for this integral:

$$n \int_0^{n^{-(1+\beta)}} \left(\log \frac{1}{t} \right) A t^\beta\, dt + \log n^{(1+\beta)}$$

$$\le Cnt^{1+\beta}\Big|_{t=n^{-(1+\beta)}} \log n^{(1+\beta)} + (1 + \beta) \log n \le C \log n.$$

This shows that the second term is at most $C \log Ln$, and similar estimates hold for the first and third terms. Thus, altogether we have

$$\sum_2 \le C \log Ln \le CL \log n.$$

Now let us consider $\sum_3$. For $j > j_0$ we set

$$A_j = |I_{j_0+1}| + \cdots + |I_j|.$$

Then for $t \in I_j$ we get from (9.9)

$$\log \left| \frac{x - \xi_j + iLm_j}{x - t} \right| \le \log \frac{|A_j + iCL|I_j||}{A_{j-1}}$$

$$= \log \frac{A_j}{A_{j-1}} + \frac{1}{2} \log \left(1 + C^2 L^2 \frac{|I_j|^2}{A_j^2} \right).$$

Thus,

$$\sum_3 \le \sum_{j>j_0+1} \log \frac{A_j}{A_{j-1}} + CL^2 \sum_{j>j_0+1} \frac{|I_j|^2}{A_j^2} = \sum_{31} + \sum_{32}.$$

Here the first sum is

(9.13) $$\sum_{31} = \log \frac{A_{n-1}}{A_{j_0+1}} \le \log \frac{1}{A_{j_0+1}} \le C \log n.$$

For the second one we get

(9.14) $$\sum_{32} \le CL^2 \sum_{j>j_0+1} \frac{|I_j|}{\sum_{j_0<l\le j}|I_l|} \le CL^2 \int_{|I_{j_0+1}|}^1 \frac{1}{t} \, dt \le CL^2 \log n.$$

These prove

$$\sum_3 \le CL^2 \log n.$$

The estimate of $\sum_1$ is completely analogous if we set for a $j < j_0 - 1$

$$A_j = |I_j| + \cdots + |I_{j_0-1}|,$$

and use for $t \in I_j$

$$\log \left| \frac{x - \xi_j + iLm_j}{x - t} \right| \le \log \frac{|A_j + iCL|I_j||}{A_{j+1}}.$$

Thus, we obtain in this case the upper bound $CL^2 \log n$ in (9.3).

Case II. $x \in [n^{-\tau_0}, 1/2]$. Let again $x \in I_{j_0}$. Let M_0 be the smallest index j for which $I_j \not\subseteq (0, x/2)$ and M_1 the largest one for which $I_j \not\subseteq (2x, 1]$. We split the sum in (9.12) into three ranges:

$$\sum_{j\le M_0} + \sum_{j=M_0+1}^{M_1-1} + \sum_{j\ge M_1} = \sum_1 + \sum_2 + \sum_3.$$

The estimation of the first and third terms is completely analogous to the estimate of $\sum_3$ in Case I above. In fact, consider e.g. $\sum_3$. Proceeding as in Case I we get

$$\sum_3 \le \sum_{j\ge M_1} \log \frac{A_j}{A_{j-1}} + CL^2 \sum_{j\ge M_1} \frac{|I_j|^2}{A_j^2} = \sum_{31} + \sum_{32}.$$

Now the analogue of (9.13) is

$$\sum_{31} = \log \frac{A_{n-1}}{A_{M_1-1}} \le \log \frac{1}{A_{M_1-1}} \le \log \frac{4}{x},$$

while that of (9.14) is

$$\sum\nolimits_{32} \leq CL^2 \sum_{j \geq M_1} \frac{|I_j|}{\sum_{j_0 < l \leq j} |I_l|} \leq CL^2 \int_{AM_1 - 1}^1 \frac{1}{t} dt \leq CL^2 \log \frac{2}{x}.$$

These prove

$$\sum\nolimits_3 \leq CL^2 \log \frac{1}{x},$$

and the estimate of $\sum_1$ is analogous.

Now let us consider $\sum_2$. We write with $\delta = 2(\beta - \beta_0)$

$$\sum\nolimits_2 = \sum\nolimits_{21} + \sum\nolimits_{22} + \sum\nolimits_{23},$$

where the range of the summations in $\sum_{21}$ is $j = j_0 \pm 1$, in $\sum_{22}$ it is $1 < |j - j_0| \leq x^{-\delta}$, while in $\sum_{23}$ it is $x^{-\delta} \leq |j - j_0|$, and recall that in every case we have $I_j \subseteq (x/2, 2x)$. In $\sum_{22}$ we can utilize once more the technique of (9.13) and (9.14). In fact, if M_2 is the largest integer that is smaller than $x^{-\delta}$, then the part $\sum_{22}'$ of $\sum_{22}$ which corresponds to the indices $j_0 + 1 < j \leq j_0 + M_2$ can be estimated with the method of (9.13) and (9.14) as

$$\begin{aligned}
\sum\nolimits_{23}' &\leq \log \frac{A_{j_0+M_2}}{A_{j_0+1}} + CL^2 \int_{|I_{j_0+1}|}^{A_{j_0+M_2}} \frac{1}{t} dt \\
&\leq CL^2 \int_{|I_{j_0+1}|}^{A_{j_0+M_2}} \frac{1}{t} dt = CL^2 \log \frac{A_{j_0+M_2}}{A_{j_0+1}}.
\end{aligned}$$

Here we can use (9.11) according to which

$$A_{j_0+M_2} \leq M_2 A(x/4)^{-\beta_0}/n \leq A(x/4)^{-\beta_0} x^{-\delta}/n$$

and

$$A_{j_0+1} \geq \frac{1}{A} (4x)^{-\beta}/n,$$

These imply via the preceding formula

$$\sum\nolimits_{23}' \leq CL^2 \log \frac{1}{x}.$$

The other part of $\sum_{23}$ (corresponding to the indices $j_0 - M_2 \leq j < j_0 - 1$) can be similarly handled and we arrive at

$$\sum\nolimits_{23} \leq CL^2 \log \frac{1}{x}.$$

In $\sum_{21}$ we have three terms. As before, we get for the middle one

$$n \int_{I_{j_0}} \log \left| \frac{x - \xi_{j_0} + iLm_{j_0}}{x - t} \right| u_n(t) \, dt \leq n \int_{I_{j_0}} \log \frac{||I_{j_0}| + iCL|I_{j_0}||}{|x - t|} u_n(t) \, dt$$

$$\leq \frac{1}{2}\log((1 + C^2 L^2)) + n \int_{I_{j_0}} \log \frac{|I_{j_0}|}{|x - t|} u_n(t)\, dt.$$

Since the integral of u_n on the interval I_j is $1/n$, there must be a point $t_1 \in I_j$ such that $u_n(t_1) \leq 1/n|I_j|$. But for any $t \in I_j$, $I_j \subseteq (x/2, 2x)$, the ratio $u_n(t)/u_n(t_1)$ is less than $CA^2 x^{\beta - \beta_0}$ (see (9.1)–(9.2)), hence for any $t \in I_j$ we have $n u_n(t) \leq CA^2 x^{\beta - \beta_0}/|I_j|$. Thus, if we break the last integral into two parts, for integration over $|x - t| \leq |I_j| x^\delta$ and the rest, then we can deduce the following upper bound for this integral:

$$\delta \log \frac{1}{x} + \frac{CA^2 x^{\beta - \beta_0}}{|I_j|} \int_{|x-t| \leq |I_j| x^\delta} \log \frac{|I_{j_0}|}{|x - t|}\, dt \leq$$
$$\delta \log \frac{1}{x} + CA^2 x^{\beta - \beta_0} x^\delta \log \frac{1}{x}.$$

This shows that the second term in $\sum_{21}$ is at most $CL^2 \log 1/x$ (recall the choice of δ: $\delta = 2(\beta_0 - \beta) > \beta_0 - \beta$), and similar estimates hold for the first and third terms. Thus, altogether we have

$$\sum_{21} \leq CL^2 \log \frac{1}{x}.$$

Thus it has left to deal with the sum $\sum_{23}$. For the j's in $\sum_{23}$ we shall make use of the inequality: if $\eta \geq -1/4$ or if $|\theta| \geq |\eta|$, then

$$(9.16) \quad \left| \log |1 + \eta + i\theta| - \eta \right| = \left| \frac{1}{2}\log(1 + 2\eta + \eta^2 + \theta^2) - \eta \right| \leq 2(\eta^2 + \theta^2).$$

When $t \in I_j$ then $|t - \xi_j| \leq |I_j| \leq Lm_j$, and so by the preceding inequality

$$\log \left| \frac{x - \xi_j + iLm_j}{x - t} \right| = \log \left| 1 + \frac{t - \xi_j + iLm_j}{x - t} \right| \leq \frac{t - \xi_j}{x - t} + 2\frac{|t - \xi_j|^2 + L^2 m_j^2}{(x - t)^2}.$$

The first term on the right is

$$\frac{t - \xi_j}{x - \xi_j} + \frac{(t - \xi_j)^2}{(x - t)(x - \xi_j)} = \frac{t - \xi_j}{x - \xi_j} + O(|I_j|^2 / \mathrm{dist}(x, I_j)^2).$$

Thus, if we multiply the last but one inequality with $n u_n(t)$ and integrate on I_j, then by taking into account the fact that the integral of the term

$$\frac{t - \xi_j}{x - \xi_j} u_n(t)$$

vanishes because of the choice of ξ_j, we finally get that

$$\sum_{23} \leq CL^2 \sum_{|j - j_0| > M_2,\ I_j \subseteq (x/2, 2x)} \frac{|I_j|^2}{\mathrm{dist}(x, I_j)^2}.$$

Taking into account that by (9.11) here

$$|I_j| \le A(x/4)^{-\beta_0}/n$$

while

$$\text{dist}(x, I_j) \ge A_{j-1} \ge (j - j_0 - 1)\frac{1}{A}(4x)^{-\beta}/n,$$

we can continue the last estimate as

$$\sum_{23} \le CL^2 A^2 x^{2(\beta - \beta_0)} \sum_{|j - j_0| > M_2} \frac{1}{(j - j_0)^2} \le CL^2 A^2 x^{2(\beta - \beta_0)} M_2^{-1}$$

$$\le CL^2 A^2 x^{2(\beta - \beta_0)} x^\delta = CL^2$$

by the choice of $\delta = 2(\beta_0 - \beta)$.

The inequalities that we have proved so far verify the upper bound in (9.3) for $x \ge n^{-\tau_0}$.

With this the proof of the upper estimate in (9.3) is complete.

The asymptotic estimate

Let $x \ge L^{-6}$. We are going to verify (9.4).

Let $j = 0, \ldots, N_1 - 1$ be those indices for which

$$I_j \subseteq [0, L^{-8}],$$

and let I be the union of the rest of the I_j's:

$$I = \cup_{j=N_1}^{n-1} I_j.$$

Let u_n^* be the restriction of u_n to I, and similarly, let

$$Q_n^*(x) = \prod_{j=N_1}^{n-1} (x - \xi_j + iLm_j),$$

i.e. in Q_n^* we only keep those terms from Q_n that correspond to indices for which I_j is lying in I.

First we remark that

$$0 \le \left(\log|Q_n(x)| + nU^{u_n}(x)\right) - \left(\log|Q_n^*(x)| + nU^{u_n^*}(x)\right)$$

$$= \sum_{j=0}^{N_1-1} \int_{I_j} \log\left|\frac{x - \xi_j + iLm_j}{x - t}\right| u_n(t)\, dt.$$

Here the terms under the integral sign are bounded by

$$\log\left|1 + \frac{t - \xi_j + iLm_j}{x - t}\right| \le \frac{t - \xi_j}{x - t} + C\frac{(1 + L^2)|I_j|^2}{\text{dist}(x, I_j)^2} \le \frac{t - \xi_j}{x - \xi_j} + C\frac{(1 + L^2)|I_j|^2}{\text{dist}(x, I_j)^2},$$

where we used (9.16). The integral against $u_n(t)\,dt$ of the first term on the right vanishes again because of the choice of ξ_j, hence,

$$(9.17) \qquad 0 \le \left(\log|Q_n(x)| + nU^{u_n}(x)\right) - \left(\log|Q_n^*(x)| + nU^{u_n^*}(x)\right)$$

$$\le CL^2 \sum_{j=0}^{N_1-1} \frac{|I_j|^2}{\operatorname{dist}(x,I_j)^2} \le CL^2 \left(\sum_{j=0}^{N_1-1} \frac{|I_j|}{2/x}\right)^2 \le CL^2 \left(\frac{L^{-8}}{L^{-6}}\right)^2 \le \frac{C}{L^2}$$

uniformly in $x \in [L^{-6}, 1/2]$.

Thus, it has left to consider

$$\log|Q_n^*(x)| + nU^{u_n^*}(x).$$

Fix an $x_0 \in [L^{-6}, 1/2]$ and consider the functions

$$U_n(x) = \int_I \log\left|x - t + \frac{iL}{nv_n(t)}\right| u_n(t)\,dt$$

and

$$U_{n,x_0}(x) = \int_I \log\left|x - t + \frac{iL}{nv_n(x_0)}\right| u_n(t)\,dt.$$

Seeing that now x_0 is fixed and the weights u_n are uniformly equicontinuous, the argument applied in Section 2 gives that

$$n\left(U_{n,x_0}(x) + U^{u_n^*}(x)\right) = \pi L \frac{u_n(x)}{v_n(x_0)}(1 + o(1))$$

uniformly in $x \in [L^{-6}, 1/2]$, and it is also easy to see that this limit relation uniformly holds also in $x_0 \in [L^{-6}, 1/2]$. In particular, we have

$$(9.18) \qquad n\left(U_{n,x_0}(x_0) + U^{u_n^*}(x_0)\right) = \pi L \frac{u_n(x_0)}{v_n(x_0)}(1 + o(1))$$

uniformly in $x_0 \in [L^{-6}, 1/2]$.

Now let us consider

$$n\left(U_n(x_0) - U_{n,x_0}(x_0)\right) = n\int_I \log\left|1 + \frac{iL}{n}\frac{v_n(t)^{-1} - v_n(x_0)^{-1}}{x_0 - t + iLv_n(x_0)^{-1}n^{-1}}\right| u_n(t)\,dt.$$

Since here

$$\left|\frac{v_n(t)^{-1} - v_n(x_0)^{-1}}{x_0 - t + iLv_n(x_0)^{-1}n^{-1}}\right| \le K_L$$

independently of $x_0 \in [L^{-6}, 1/2]$ and n, for large enough n we get

$$n\log\left|1 + \frac{iL}{n}\frac{v_n(t)^{-1} - v_n(x_0)^{-1}}{x_0 - t + iLv_n(x_0)^{-1}n^{-1}}\right| \le 2K_L.$$

This and the limit

$$\lim_{n \to \infty} n \log \left| 1 + \frac{iL}{n} \frac{v_n(t)^{-1} - v_n(x_0)^{-1}}{x_0 - t + iLv_n(x_0)^{-1}n^{-1}} \right| = 0$$

imply via the dominated convergence theorem that

$$\lim_{n \to \infty} n \left(U_n(x_0) - U_{n,x_0}(x_0) \right) = 0$$

uniformly in $x_0 \in [L^{-6}, 1/2]$.

Thus, altogether we get from the last limit and (9.18) that

(9.19) $$n \left(U_n(x) + U^{u_n^*}(x) \right) = \pi L \frac{u_n(x)}{v_n(x)} (1 + o(1))$$

uniformly in $x \in [L^{-6}, 1/2]$.

Note that the functions

(9.20) $$g_n(x) := \pi \frac{u_n(x)}{v_n(x)}$$

on the right are lying in between $\pi/2$ and 2π for $n \geq n_L$ (see (9.6) and the definition of v_n in (9.5)), and they are uniformly equicontinuous on every closed subinterval of $(0, 1)$ (in particular, on $[L^{-6}, 1/2]$) by our assumption concerning the functions u_n. From our considerations it will be clear that these are the functions g_n in the lemma.

Next we estimate the sum

(9.21) $$\log |Q_n^*(x)| - nU_n(x)$$

for $x \in [L^{-6}, 1/2]$. Let $x \in I_{j_0}$. Then j_0 is a large index (larger than const$\cdot n$) in view of the fact that by (9.2) we have $v_n(t) \geq L^{-8\beta_0}$ if $\in [L^{-7}, L^{-6}]$, hence there are at least $c_L n$, $c_L > 0$ intervals I_j lying in $\in [L^{-7}, L^{-6}]$. We can write this difference in the form

$$\left| \log |Q_n^*(x)| - nU_n(x) \right| = \left| \sum_{j=N_1}^{n-1} n \int_{I_j} \log \left| \frac{x - t + iLv_n(t)^{-1}n^{-1}}{x - \xi_j + iLm_j} \right| u_n(t) \, dt \right|$$

$$\leq \left| \left(\sum_{j=N_1}^{j_0-L} + \sum_{j=j_0-L+1}^{j_0+L-1} + \sum_{j=j_0+L}^{n-1} \right) n \int_{I_j} \log \left| \frac{x - t + iLv_n(t)^{-1}n^{-1}}{x - \xi_j + iLm_j} \right| u_n(t) \, dt \right|$$

$$= \left| \sum_1 \right| + \left| \sum_2 \right| + \left| \sum_3 \right|.$$

First we estimate $\sum_2$. The definition of m_j (see (9.7)) shows that for the j's appearing in the sum $\sum_2$ there is a $\zeta_j \in I_j$ such that $v_n(\zeta_j)^{-1} = nm_j$, and this easily implies that for $t \in I_j$

$$(9.22) \qquad |v_n(t)^{-1}n^{-1} - m_j| = \frac{1}{n}|v_n(t)^{-1} - v(\zeta_j)^{-1}|$$

$$\leq C\frac{|I_j|}{n}\frac{|v_n'(\xi_j)|}{v_n(\xi_j)^2} \leq C_L\frac{m_j}{n}.$$

Thus

$$\left|\sum_2\right| \leq 2(L+1)\frac{1}{2}\log\frac{|I_j|^2 + L^2m_j^2(1+O(n^{-1}))}{L^2m_j^2} \leq \frac{C}{L},$$

where we have also used that by (9.9) $m_j \sim |I_j|$.

The sum $|\sum_3|$ can be written in the form

$$\left|\sum_3\right| = \left|\sum_{j=j_0+L}^{n-1} n \int_{I_j} \log\left|1 + \frac{t - \xi_j + iL(v_n(t)^{-1}n^{-1} - m_j)}{x - \xi_j + iLm_j}\right| u_n(t)\,dt\right|.$$

We would like to bring the estimate (9.16) into play, so first we have to show the lower bound $-1/4$ for the real part of the ratio standing after 1:

$$\Re\frac{t - \xi_j + iL(v_n(t)^{-1}n^{-1} - m_j)}{x - \xi_j + iLm_j}$$

$$= \frac{(t - \xi_j)(x - \xi_j) + L^2(v_n(t)^{-1}n^{-1} - m_j)m_j}{(x - \xi_j)^2 + L^2m_j^2} \geq -\frac{|I_j|}{\xi_j - x} - \frac{C_L}{n} \geq -\frac{1}{4},$$

provided L is sufficiently large (depending on A and β), and n is sufficiently large depending on L. At the last but one step we used (9.22) and at the last step we used that $\xi_j \in I_j$ with index $j \geq j_0 + L$, hence for large L we have $-|I_j|/(\xi_j - x) \geq -1/8$ by the uniform equicontinuity of the functions u_n on compact subsets of $(0, 1)$.

Thus, we can apply (9.16) to the integrands in the sums in $\sum_3$ and we obtain

$$\left|\sum_3 - \sum_{j=j_0+L}^{n-1} n \int_{I_j} \Re\frac{t - \xi_j + iL(v_n(t)^{-1}n^{-1} - m_j)}{x - \xi_j + iLm_j}u_n(t)\,dt\right|$$

$$\leq 2\sum_{j=j_0+L}^{n-1} \frac{|I_j|^2 + C_LL^2(m_j/n)^2}{(x - \xi_j)^2 + L^2m_j^2},$$

where we have used again (9.22). By the definition of the numbers ξ_j and m_j each integral

$$\int_{I_j} \Re\frac{t - \xi_j + iL(v_n(t)^{-1}n^{-1} - m_j)}{x - \xi_j + iLm_j}u_n(t)\,dt$$

$$= \left(\Re\frac{1}{x - \xi_j + iLm_j}\right)\int_{I_j}(t - \xi_j)u_n(t)\,dt -$$

$$\Im\left(\frac{1}{x - \xi_j + iLm_j}\right)\int_{I_j}(v_n(t)^{-1}n^{-1} - m_j)u_n(t)\,dt$$

vanishes, hence we can continue the above inequality as

$$\left|\sum_3\right| \le 2 \sum_{j \ge j_0+L} \frac{|I_j|^2 + C_L L^2 |I_j|^2 n^{-2}}{\text{dist}(x, I_j)^2} \le 3 \sum_{j \ge j_0+L} \frac{|I_j|^2}{\text{dist}(x, I_j)^2}$$

if n is sufficiently large. Now we write the last sum as $\sum_{31} + \sum_{32}$ where in $\sum_{31}$ we sum for all indices $j \ge j_0 + L$ for which $I_j \subseteq (x, x + d_L)$ with the constant d_L from (9.6). The choice of d_L yields for $I_j, I_l \subseteq (x, x + d_L)$

$$\frac{1}{2} \le |I_j|/|I_l| \le 2,$$

hence

$$\sum_{31} \le C \sum_{j_0+L}^{\infty} \frac{1}{(j - j_0)^2} \le \frac{C}{L}.$$

For the second sum we get from (9.10)

$$\sum_{32} \le \frac{\max_j |I_j|}{d_L^2} \sum |I_j| \le C n^{-r_0} d_L^{-2} \le L^{-1}$$

if n is sufficiently large.

The estimate of $|\sum_1|$ is similar:

$$\sum_1 \le 2 \sum_{j=N_1}^{j_0-L} \frac{|I_j|^2 + C_L L^2 |I_j|^2 n^{-2}}{\text{dist}(x, I_j)^2} \le 3 \sum_{j=N_1}^{j_0-L} \frac{|I_j|^2}{\text{dist}(x, I_j)^2}$$

$$\le C \sum_{j \le j_0-L} \frac{1}{(j - j_0)^2} + \frac{\max_j |I_j|}{d_L^2} \sum |I_j| \le C L^{-1}.$$

Our estimates from (9.21) show that for $n \ge n_L$

(9.23) $$\left| \log |Q_n^*(x)| - n U_n(x) \right| \le \frac{C}{L}$$

for $x \in [L^{-6}, 1/2]$ with a constant C independent of L provided L is sufficiently large, say $L \ge L_0$.

Finally, from (9.17), (9.19) and (9.23) we obtain (9.4) in our lemma on the interval $[L^{-6}, 1/2]$ with the functions (9.20).

It is obvious what modifications we have to make to cover the interval $[1/2, 1]$, as well. Practically the only necessary change is to choose now

$$m_j = |I_j| \qquad \text{if} \ \ I_j \subseteq [0, L^{-9}] \cup [1 - L^{-9}, 1],$$

otherwise let

$$m_j = \int_{I_j} u_n(t)/v_n(t) \, dt.$$

With this change the proof goes over with trivial modifications.   ∎

Closer inspection of the above proof shows that the estimates in Lemma 9.1 hold in larger ranges:

$$(9.24) \qquad 0 \le \log|Q_n(x)| + nU^{u_n}(x)$$

$$\le \begin{cases} BL^2 \log n & \text{if } \operatorname{dist}(x, \{0,1\}) \le n^{-\tau} \\ BL^2 \log\left(1 + 1/\operatorname{dist}(x, \{0,1\})\right) & \text{for all other } x, \end{cases}$$

and with some continuous functions g_n that are uniformly bounded and uniformly equicontinuous on every compact subset of $(0,1)$ and vanish outside $[0,1]$

$$(9.25) \qquad \left|\log|Q_n(x)| + nU^{u_n}(x) - Lg_n(x)\right| \le BL^{-1}$$

if $\operatorname{dist}(x, \{0,1\}) \ge L^{-6}$. This remark allows us to extend Lemma 9.1 to several intervals or to have zero or infinite singularities of the weights u_n inside their support. In fact, let us assume that (a_k, b_k) are finitely many disjoint intervals with possibly common endpoints. We set $\Sigma_0 = \cup(a_k, b_k)$ and $X = \{a_k, b_k\}$. Suppose that $\{u_n\}$ is a sequence of nonnegative functions on Σ_0 satisfying the following conditions: $\{u_n\}$ is uniformly equicontinuous on every compact subinterval of Σ_0,

$$\int_{\Sigma_0} u_n = 1,$$

and otherwise on every (a_k, b_k) the functions u_n satisfy inequalities that are analogous to (9.1)—(9.2) with some β_j, β_{0j}, A and $\tau > 0$. Then there is an $L_0 > 1$ such that for every $L > L_0$ there are polynomials Q_n of degree at most n such that for large n, say $n \ge n_L$

$$(9.26) \quad -C \;\le\; \log|Q_n(x)| + nU^{u_n}(x)$$

$$\le \begin{cases} BL^2 \log n & \text{if } \operatorname{dist}(x, X) \le n^{-1} \\ BL^2 \log\left(1 + 1/\operatorname{dist}(x, X)\right) & \text{for all other } x, \end{cases}$$

and with some continuous functions g_n that are uniformly bounded and uniformly equicontinuous on every compact subset of Σ_0 and vanish outside the closure of Σ_0,

$$\left|\log|Q_n(x)| + nU^{u_n}(x) - Lg_n(x)\right| \le BL^{-1} \qquad \text{if } \operatorname{dist}(x, X) \ge L^{-6}.$$

Here B is an absolute constant depending only on the parameters in the assumptions on the u_n's. The proof can be based on Lemma 9.1. In fact, if for an n the function u_n has integral of the form $l_{k,n}/n$ with some integer $l_{k,n}$ on each (a_k, b_k), then we can construct the polynomials in Lemma 9.1 for the weight function $u_n\big|_{(a_k, b_k)}$ with the modification that now the degree has to be $l_{k,n}$ (to match the integral of $nu_n\big|_{(a_k, b_k)}$). Then by the above remarks the product of these polynomials will be appropriate (actually with $C = 0$ on the left of (9.26)). In general, however, the integral

$$\alpha_{k,n} := \int u_n \Big|_{(a_k, b_k)}$$

is not of the form $l_{k,n}/n$. In that case let us fix for each k a nonnegative continuous function s_k with compact support in (a_k, b_k) and with integral 1, and consider the weights

$$u_n^*(t) \quad := \quad u_n(t) - \sum_{k>1} \left(\alpha_{k,n} - \frac{1}{n}[n\alpha_{k,n}] \right) s_k(t)$$

$$+ \quad \left(\sum_{k>1} (\alpha_{k,n} - \frac{1}{n}[n\alpha_{k,n}]) \right) s_1(t).$$

These satisfy the assumptions that their integral on any (a_k, b_k) is of the form $l_{k,n}/n$, so for these there are polynomials Q_n with the required properties. Finally, all we have to mention is that the functions

$$n(U^{u_n}(x) - U^{u_n^*}(x))$$

are uniformly bounded and equicontinuos on the real line. Hence the above Q_n's are suitable for the u_n's, as well (this is where we need possibly zonzero C on the left of (9.26)).

We could have even allowed the set Σ_0 vary with n, but we do not pursue this direction any further.

10   Approximation in geometric means

In this section we prove two theorems that are in some sense refinements of the theorems in Part 2 in the special case considered here.

Theorem 10.1 *Suppose that $\{v_n\}$ is a sequence of nonnegative functions on $(0,1)$ satisfying the following conditions: $\{v_n\}$ is uniformly equicontinuous on every compact subinterval of $(0,1)$,*

$$\int_0^1 v_n = 1,$$

and for some constants A, $\beta > -1$ and β_0

(10.1) $v_n(t) \leq A(t(1-t))^\beta,$ $t \in (0,1),$

(10.2) $u_n(t) \geq \dfrac{1}{A}(t(1-t))^{\beta_0},$ $t \in (0,1),$

and set

$$w_n(x) := \exp(U^{v_n}(x)).$$

If $1 \geq \gamma \geq 0$ and $u(x)$ is any positive continuous function on $[0, 1]$, then there are polynomials H_n of degree at most n such that for

$$(10.3) \qquad h_n(x) = w_n^n(x)|H_n(x)|(x(1-x))^{-\gamma}u(x)$$

we have

$$h_n(x) \geq 1, \qquad x \in (0, 1),$$

$$(10.4) \qquad \lim_{n \to \infty} h_n(x) = 1$$

uniformly on compact subsets of $(0, 1)$, and

$$(10.5) \qquad \lim_{n \to \infty} \int_0^1 \frac{\log h_n(x)}{\sqrt{x(1-x)}} \, dx = 0.$$

It will be of utmost importance that *the degree of H_n can be somewhat smaller than n, namely we can have* $\deg(H_n) = n - i_n$ *where* $i_n \to \infty$. This will follow from the proof (with $i_n = n^\delta$ for some $\delta > 0$).

Here, exactly as in Lemma 9.1, we actually need the inequality in (10.2) only for $t \in [n^{-\tau}, 1 - n^{-\tau}]$.

Proof. For some $\rho > 0$ let

$$\lambda_n = 1 + n^{-\rho},$$

and consider the weights

$$w_n^* := w_n^{\lambda_n}$$

and the corresponding weighted energy problem of Section 4.1 with extremal measure μ_n^* and $S_n^* = \text{supp}(\mu_n^*)$.

We shall again concentrate on the behavior around the left endpoint $x = 0$.

Fix a C_0 and set $a_n = C_0 n^{-\rho/(\beta_0 + 1)}$. Since for $x \in [a_n/2, 1/2]$ we have on the interval $[x/2, 3x/2]$ the inequality

$$v_n(y) \geq \frac{1}{A \cdot 4^{\beta_0}} y^{\beta_0},$$

and since

$$\frac{1}{A \cdot 4^{\beta_0}} y^{\beta_0} \frac{x}{2} \geq \frac{C_0^{1+\beta_0}}{A \cdot 8^{\beta_0+1}} n^{-\rho},$$

we get from Lemma 5.8 that for large enough C_0 the interval $[a_n/2, 1/2]$ is part of S_n^*.

Let $J_n = [a_n/2, 3/4]$. Lemma 5.7 can be applied to deduce

$$(10.6) \qquad d\mu_n^*(t) \leq \lambda_n v_n(t) \, dt \leq 2At^\beta \, dt$$

for all $t \in (0, 1/2]$. In particular, each μ_n^* is absolutely continuous.

By Lemma 5.7 we also get that

$$d\mu^*(t) \geq \lambda_n v_n \, dt - (\lambda_n - 1)d\mu_{J_n}(t), \qquad t \in [a_n/2, 3/4],$$

where the last measure

$$d\mu_{J_n}(t) = \frac{1}{\pi} \frac{1}{\sqrt{(t - a_n/2)(3/4 - t)}} \, dt$$

on the right is the equilibrium measure of the interval J_n. Thus, on $[a_n, 1/2]$ we get

$$(10.7) \qquad (\mu_n^*(t))' \geq v_n(t) - n^{-\rho}(t - a_n/2)^{-1/2} \geq \frac{1}{2} v_n(t)$$

for large enough n and $t \geq C_0 n^{-\rho/(1+\beta_0)}$ which follows from (10.2) and the choice of the numbers a_n.

Lemma 5.7 together with the assumptions on v_n also show that the functions

$$(10.8) \qquad u_{[n/\lambda_n]}(t) := (\mu_n^*(t))'$$

are uniformly equicontinuous on compact subintervals of $(0, 1/2]$.

Thus, in summary we can say that $\{u_n\}$ satisfy on the interval $(0, 1/2]$ the assumptions of Lemma 9.1 with $\tau = \rho/2(1+\beta_0)$ where we consider (9.2) only for $t \in [n^{-\tau}, 1 - n^{-\tau}]$ (here the factor $1/2$ is put into the expression of $\tau = \rho/2(1+\beta_0)$ to get rid of the constant C_0).

The same analysis can be done on the interval $[1/2, 1)$, and we can conclude that the densities $\{u_n\}$ from (10.8) satisfy all the assumptions of Lemma 9.1 (with $\tau = \rho/2(1 + \beta_0)$) for large enough n.

On applying Lemma 9.1 to $u_{[n/\lambda_n]}$ and to $[n/\lambda_n]$ rather than to n we get that for every $L \geq L_0$ there is a sequence of polynomials $Q_{[n/\lambda_n]}$ of degree at most $[n/\lambda_n]$ such that for sufficiently large n $(n \geq n_L)$ we have

$$(10.9) \qquad w_n(x)^{\lambda_n[n/\lambda_n]} |Q_{[n/\lambda_n]}(x)| \geq 1$$

for every $x \in (0, 1)$,

$$(10.10) \qquad w_n(x)^{\lambda_n[n/\lambda_n]} |Q_{[n/\lambda_n]}(x)| = e^{Lg_n(x) + O(L^{-1})}$$

for $L^{-6} \leq x \leq 1 - L^{-6}$,

$$(10.11) \qquad w_n(x)^{\lambda_n[n/\lambda_n]} |Q_{[n/\lambda_n]}(x)| = \left(\frac{1}{x(1-x)}\right)^{O(L^2)}$$

for $n^{-\tau} \leq x \leq L^{-6}$ or $n^{-\tau} \leq 1 - x \leq L^{-6}$, and

$$(10.12) \qquad w_n(x)^{\lambda_n[n/\lambda_n]} |Q_{[n/\lambda_n]}(x)| = n^{O(L^2)}$$

for $0 \leq x \leq n^{-\tau}$ or $0 \leq 1 - x \leq n^{-\tau}$, where the O is uniform in L and $n \geq n_L$ and the functions $g_n \geq 0$ are uniformly bounded and uniformly equicontinuous on compact subsets of $(0, 1)$.

The weights $w_n^{\lambda_n[n/\lambda_n]}$ and w_n^n differ only in a multiplicative factor $w_n^{\delta_n}$, $0 \leq \delta_n := n - \lambda_n[n/\lambda_n] \leq 1$ which are also uniformly bounded on $(0, 1)$ (c.f.

the assumptions of the theorem concerning v_n) and uniformly equicontinuous on compact subsets of $(0, 1)$.

It is now easy to find for every $q > 0$ polynomials $R_{n-[n/\lambda_n]}$ of degree at most $n - [n/\lambda_n]$ such that for sufficiently large n the following estimates hold with some absolute constant C_L independent of q and n:

$$\lim_{n \to \infty} R_{n-[n/\lambda_n]}(x) e^{Lg_n(x)} w_n^{\delta_n}(x) = \frac{(x(1-x))^\gamma}{u(x)}$$

uniformly on $[L^{-6} + q, 1 - L^{-6} - q]$,

$$1 \le R_{n-[n/\lambda_n]}(x) e^{Lg_n(x)} w_n^{\delta_n}(x) \le C_L$$

on the intervals $[L^{-6}, L^{-6} + q]$ and $[1 - L^{-6} - q, 1 - L^{-6}]$, and

$$1 \le R_{n-[n/\lambda_n]}(x) \le 2$$

on $[0, L^{-6}]$ and on $[1 - L^{-6}, 1]$. In fact, the functions

$$r_n(x) =$$

$$
\begin{cases}
e^{-Lg_n(x)} w_n^{-\delta_n}(x)(x(1-x))^\gamma / u(x) & \text{if } x \in [L^{-6} + q, 1 - L^{-6} - q] \\
1 & \text{if } x \in [0, L^{-6}] \cup [1 - L^{-6}, 1] \\
\text{linear} & \text{on } [L^{-6}, L^{-6} + q] \text{ and on} \\
& \quad [1 - L^{-6} - q, 1 - L^{-6}]
\end{cases}
$$

form a compact subset of $C[0, 1]$ (they are uniformly bounded and uniformly equicontinuous in n), hence for every $\epsilon > 0$ there is an m such that for every n there are polynomials $R_{m,n}^*$ of degree at most m such that

$$e_{m,n} := \|r_n - R_{m,n}^*\|_{C[0,1]} \le \epsilon.$$

This means that if we choose $R_{n-[n/\lambda_n]}$ as

$$R_{n-[n/\lambda_n]}(x) := R_{n-[n/\lambda_n],n}^*(x) + \epsilon_{n-[n/\lambda_n],n} + e^{-C_1/L},$$

with some appropriate but fixed C_1 (that depends only on the constants in the preceding inequalities), then (by $n - [n/\lambda_n] \ge n^\rho/2$) this choice will satisfy the above inequalities.

Thus, althogether we get for the polynomials

$$H_n(x) = Q_{[n/\lambda_n]}(x) R_{n-[n/\lambda_n]}(x)$$

of degree at most n for every large n, say $n \ge n_L$ the estimates

$$1 \le w_n^n(x)|H_n(x)|(x(1-x))^{-\gamma} u(x) \le e^{CL^{-1}}$$

for $L^{-6} + q \le x \le 1 - L^{-6} - q$,

$$1 \le w_n^n(x)|H_n(x)|(x(1-x))^{-\gamma} u(x) \le C \cdot C_L$$

for $L^{-6} \leq x \leq L^{-6} + q$ and $L^{-6} \leq 1 - x \leq L^{-6} + q$,

$$1 \leq w_n^n(x)|H_n(x)| \leq \left(\frac{1}{x(1-x)}\right)^{CL^2}$$

for $n^{-\tau} \leq x \leq L^{-6}$ and $n^{-\tau} \leq 1 - x \leq L^{-6}$, and

$$1 \leq w_n^n(x)|H_n(x)| \leq n^{CL^2}$$

for $0 \leq x \leq n^{-\tau}$ or $0 \leq 1 - x \leq n^{-\tau}$, where the constant C is independent of q and L and $n \geq n_{L,q}$.

Since these inequalities actually give estimates on h_n (see (10.3)), and they show that $h_n \geq 1$, all we have to verify that the geometric means in (10.5) are as small as we like. The preceding estimates give for this mean and for large n the inequality

$$
\begin{aligned}
0 \; &\leq \; \int_0^1 \frac{\log h_n(x)}{\sqrt{x(1-x)}}\, dx \\
&\leq \; C\left(n^{-\tau/2}L^2 \log n + L^2 \int_0^{L^{-6}} \frac{\log 1/x}{\sqrt{x(1-x)}} + q \log C_L + L^{-1}\right).
\end{aligned}
$$

Choosing now first L large, then q small, we can make the right hand side as small as we wish. With this the proof is complete. ∎

Next we prove the following companion to Theorem 10.1.

Theorem 10.2 *Suppose that $\{v_n\}$ is a sequence of nonnegative functions on $(0, 1)$ satisfying the conditions of the preceding theorem, and*

$$w_n(x) := \exp(U^{v_n}(x)).$$

If $1/2 \geq \gamma \geq 0$ and $u(x)$ is any positive continuous function on $(0, 1)$, then there are polynomials H_{n-1} of degree at most $n - 1$ such that H_{n-1} does not vanish on $(0, 1)$, and for the function

(10.13) $$h_n(x) = w_n^n(x)|H_{n-1}(x)\sqrt{x(1-x)}|(x(1-x))^{-\gamma}u(x)$$

we have

(10.14) $$h_n(x) \leq 1, \qquad x \in (0, 1),$$

(10.15) $$\lim_{n \to \infty} h_n(x) = 1$$

uniformly on compact subsets of $(0, 1)$, and

(10.16) $$\lim_{n \to \infty} \int_0^1 \frac{\log h_n(x)}{\sqrt{x(1-x)}}\, dx = 0.$$

Note, that in the present case we need the factor $\sqrt{x(1-x)}$ in h_n when $\gamma > 0$ in order to achieve (10.14).

Again, *we can have* $\deg(H_{n-1}) = n - i_n$ *where* $i_n \to \infty$ (say $i_n = n^\delta$ with some $\delta > 0$).

Proof. Proceed as before in the proof of Theorem 10.1 until (10.9)—(10.12). Choose with a large D, to be specified below, the number $N = DL^2$, and consider

$$Q^*_{[n/\lambda_n]}(x) := Q_{[n/\lambda_n]}(x)\left(x + \frac{1}{n}\right)^N \left(1 - x + \frac{1}{n}\right)^N.$$

If D is large (depending only on the constants in (10.10)—(10.12)), then we can easily get from (10.10)—(10.12) that

$$w_n(x)^{\lambda_n[n/\lambda_n]}|Q^*_{[n/\lambda_n]}(x)| \le 1$$

for all $x \in [0,1]$,

$$w_n^{\lambda_n[n/\lambda_n]}(x)|Q^*_{[n/\lambda_n]}(x)| = e^{g^*_n(x)+O(L^{-1})}$$

for $L^{-6} \le x \le 1 - L^{-6}$,

$$w_n(x)^{\lambda_n[n/\lambda_n]}|Q^*_{[n/\lambda_n]}(x)| \ge (x(1-x))^{2DL^2}$$

for $0 \le x \le n^{-\tau}$ or $0 \le 1 - x \le n^{-\tau}$, where the O is uniform in L and $n \ge n_L$, and the functions

$$g^*_n(x) = Lg_n(x) + N\log((x + 1/n)(1 - x + 1/n))$$

are uniformly bounded and uniformly equicontinuous on $[L^{-6}, 1 - L^{-6}]$.

With the argument applied in the preceding proof we can find polynomials $R_{n-1-[n/\lambda_n]} - 2N$ of degree at most $n - 1 - [n/\lambda_n] - 2N$ such that for sufficiently large n the relations

$$\lim_{n \to \infty} R_{n-[n/\lambda_n]}(x)e^{g^*_n(x)}w_n^{\delta_n}(x)\sqrt{x(1-x)} = \frac{(x(1-x))^\gamma}{u(x)}$$

uniformly on $[L^{-6} + q, 1 - L^{-6} - q]$,

$$1 \ge R_{n-[n/\lambda_n]}(x)e^{g^*_n(x)}w_n^{\delta_n}(x) \ge \frac{1}{C_L}$$

on $[L^{-6}, L^{-6} + q]$ and $[1 - L^{-6} - q, 1 - L^{-6}]$, and

$$1 \ge R_{n-[n/\lambda_n]}(x) \ge \frac{1}{2}$$

on $[0, L^{-6}]$ and on $[1 - L^{-6}, 1]$, hold with some constant C_L independent of n and q.

We set

$$H_{n-1}(x) = Q_{[n/\lambda_n]}(x)R_{n-1-[n/\lambda_n]-2N}(x)e^{-C_1/L},$$

which has degree at most n. The estimates

$$1 \ge w_n^n(x)|H_{n-1}(x)\sqrt{x(1-x)}|(x(1-x))^{-\gamma}u(x) \ge e^{-C/L}$$

for $L^{-6} + q \leq x \leq 1 - L^{-6} - q$,

$$1 \geq w_n^n(x)|H_{n-1}(x)\sqrt{x(1-x)}|(x(1-x))^{-\gamma}u(x) \geq (x(1-x))^{2DL^2}\frac{1}{C_L}$$

for $L^{-6} \leq x \leq L^{-6} + q$ and $L^{-6} \leq 1 - x \leq L^{-6} + q$,

$$1 \geq w_n^n(x)|H_{n-1}(x)|(x(1-x))^{-\gamma}u(x) \geq (x(1-x))^{2DL^2}$$

for $0 \leq x \leq L^{-6}$ and $0 \leq 1 - x \leq L^{-6}$ hold with some constant C_L independent of q and $n \geq n_L$, and some absolute constant C.

These tell us first of all that $h_n \leq 1$ (see (10.13)), and then that

$$0 \geq \int_0^1 \frac{\log h_n(x)}{\sqrt{x(1-x)}}\,dx \geq -C\left(L^2\int_0^{L^{-6}}\frac{\log 1/x}{\sqrt{x(1-x)}}\,dx + q\log C_L + L^{-1}\right).$$

Choosing first L large, then q small we can make the right hand side as small as we wish. The proof is complete.   ∎

In order to be able to apply Theorems 10.1 and 10.2 we need convenient criteria in terms of the weight w itself (note that these theorems refer to the density function). In the rest of this section we discuss what smoothness conditions on w ensure that the assumptions of Theorems 10.1 and 10.2 are satisfied. We shall do this with conditions similar to those in Theorem 8.3.

Theorem 10.3 *Suppose that $\{w_n\}$, $w_n = \exp(-Q_n)$ is a sequence of weights such that the extremal support S_{w_n} is $[0,1]$ for all n, the functions $tQ_n'(t)$ are uniformly of class C^ϵ on $[0,1]$ for some $\epsilon > 0$. Suppose further that the functions $tQ_n'(t)$ are nondecreasing on $[0,1]$ and there are points $0 < c < d < 1$, and an $\eta > 0$ such that $dQ_n'(d) \geq cQ_n'(c)+\eta$ for all n. Then the conditions of Theorems 10.1 and 10.2 are satisfied and their conclusions hold.*

For example, the conditions of this theorem are true if all $tQ_n'(t)$ coincide with a single C^ϵ function, say if $Q_n(t) = t^\alpha$ for an $\alpha > 0$.

Proof. We use the representation (8.17), and set $g_n(t) = tQ_n'(t)$. The uniform C^ϵ property of the g_n's easily imply that the integrals

$$I_n(t) = \int_0^1 \frac{g_n(s) - g_n(t)}{s - t}\frac{1}{\sqrt{s(1-s)}}\,ds$$

are uniformly bounded by a constant multiply of $(x(1-x))^{-1/2+\epsilon/2}$, hence condition (10.1) is true.

We have already used in the proof of Theorem 8.3 that the uniform C^ϵ property of the g_n's implies that of the integrals $I_n(t)$ on compact subset of $(0,1)$ (see e.g. the Plemelj–Privalov theorem in [39, p. 46]), and this is more than the uniform equicontinuity of the densities u_n of the corresponding extremal measures μ_w.

Therefore, it has left to verify condition (10.2). But we have seen in the proof of Theorem 8.3 that the integrals $I_n(t)$ are uniformly bounded from below by a positive constant, and this implies (10.2) via the formula (8.17). ∎

We shall also use the following corollary (c.f. Theorems 10.1 and 10.2).

Corollary 10.4 *Suppose that $\{w_n\}$, $w_n = \exp(-Q_n)$ is a sequence of even weights such that the extremal support S_{w_n} is $[-1, 1]$ for all n, and on $[0, 1]$ the functions satisfy the conditions of the preceding theorem. If $1 \geq \gamma \geq 0$ and $u(x)$ is any positive continuous function on $[-1, 1]$, then there are polynomials H_n of degree at most n such that for*

$$h_n(x) = w_n^n(x)|H_n(x)|(1 - x^2)^{-\gamma}u(x)$$

we have

(10.17)
$$h_n(x) \geq 1, \qquad x \in (-1, 1),$$

$$\lim_{n \to \infty} h_n(x) = 1$$

uniformly on compact subsets of $(-1, 1)$, and

$$\lim_{n \to \infty} \int_0^1 \frac{\log h_n(x)}{\sqrt{1 - x^2}}\, dx = 0.$$

In a similar fashion, if $\gamma \leq 1/2$ then the conclusion also holds for some H_{n-1} and

$$h_n(x) = w_n^n(x)|H_{n-1}(x)\sqrt{1 - x^2}|(1 - x^2)^{-\gamma}u(x)$$

with (10.17) replaced by

$$h_n(x) \leq 1, \qquad x \in (-1, 1).$$

Exactly as in the case of Theorems 10.1 and 10.2 the degree of H_n can be somewhat smaller than n, namely we can have $\deg(H_n) = n - i_n$ where $i_n \to \infty$.

Proof. This corollary is an immediate consequence of Theorems 10.1 and 10.2 and *the discussion made at the end of Section 9*. In fact, by using the transformation $x \to x^2$ applied in the proof of Lemma 8.5 we can see that the assumptions of Lemma 9.1 hold true on $(-1, 0)$ and on $(0, 1)$, and at the end of Section 9 we mentioned how to use these facts to conclude the statement of Lemma 9.1 for the union of these intervals. Now the proof of Theorems 10.1 and 10.2 were based on Lemma 9.1, hence the proof can be copied to yield the corollary. ∎

Part IV
Applications

In this chapter we shall briefly discuss some applications of our results. They are here for illustration. Some of these have been achieved in less generality by different authors using different techniques, we shall give proper reference at those places.

11 Fast decreasing polynomials

In this section, we shall discuss an application of the method developed in the preceding chapters.

We call polynomials P_n, $\deg P_n \leq n$, fast decreasing on $[-1, 1]$, if they attain the value 1 at $x = 0$ and decrease fast away from the origin:

$$(11.1) \qquad P_n(0) = 1, \qquad |P_n(x)| \leq e^{-\Phi(x)}, \qquad x \in [-1, 1].$$

We shall discuss the problem with what Φ and n this is possible.

The significance of such fast decreasing polynomials lies in the fact that they approximate the "Dirac delta function" as best as possible among polynomials of a given degree, hence for example these are the best polynomial kernels for convolution operators to reproduce the identity. By integration we can get from the above polynomials good polynomial approximants S_n of the signum function in the sense

$$(11.2) \qquad |\text{sign} x - S_n(x)| \leq e^{\Phi(x)}, \qquad x \in [-1, 1],$$

which in turn can be used to construct well localized "partitions of unity" (c.f. the construction on p. 156 in [14]) consisting of polynomials of a given degree n.

The problem can be formulated in two different ways: one can ask what possible decrease (i.e. what Φ) is possible for a given degree, or, alternatively, for a given Φ what is the smallest degree n for which there are polynomials with properties (11.1). Let n_Φ denote this degree. For symmetric Φ's which are increasing on $[0, 1]$ this problem was completely solved up to a constant in [15]:

Let Φ be an even function, right continuous and increasing on $[0, 1]$. Then

$$\frac{1}{6} N_\Phi \leq n_\Phi \leq 12 N_\Phi,$$

where $N_\Phi = 0$ if $\Phi(1) \leq 0$, and

$$N_\Phi = 2 \sup_{\Phi^{-1}(0) \leq x < b} \sqrt{\frac{\Phi(x)}{x^2}} + \int_b^{1/2} \frac{\Phi(x)}{x^2} dx + \sup_{1/2 \leq x < 1} \frac{\Phi(x)}{-\log(1-x)} + 1,$$

$b = \min(\Phi^{-1}(1), 1/2)$, *otherwise.*

This estimate is given for all Φ, in particular, Φ can depend on n. As an immediate corollary we get that if φ is even and increasing on $[0,1]$, then there are polynomials P_n of degree at most n satisfying

$$(11.3) \quad P_n(0) = 1, \qquad |P_n(x)| \le C e^{-cn\varphi(x)}, \quad x \in [-1,1], \quad n = 0, 1, \dots$$

for some constants $C, c > 0$ if and only if

$$\int_0^1 \frac{\varphi(u)}{u^2} du < \infty.$$

In [49] a potential theoretical method was developed for obtaining sharp results for the largest possible c in (11.3). It was shown there that if φ is even, increasing on $[0,1]$, and $\varphi(\sqrt{x})$ is concave on $[0,1]$, then there are polynomials satisfying (11.3) only if

$$c\frac{2}{\pi} \int_0^1 \frac{\varphi(t)}{t^2\sqrt{1-t^2}} dt \le 1$$

holds, and if we have here strict inequality then such polynomials do exist. This can be applied to any $\varphi(t) = c|t|^\alpha$, $\alpha \le 2$ and we obtain that there are polynomials P_n with

$$(11.4) \qquad P_n(0) = 1, \qquad |P_n(x)| \le C e^{-nc|x|^\alpha}, \quad x \in [-1,1],$$

only if $\alpha > 1$ and $c \le \sqrt{\pi}\Gamma\left(\frac{\alpha}{2}\right)/\Gamma\left(\frac{\alpha-1}{2}\right)$, and conversely, if the strict inequality holds, then the existence of polynomials with property (11.4) was proven in [49]. The existence of the polynomials in question when the equality holds has remained open (it was resolved in [31] for $\alpha = 2$).

Now the discretization method of Sections 2 and 3 enables us to settle this problem.

Theorem 11.1 *Let $\alpha \le 2$. Then there are polynomials P_n with property (11.4) if and only if $\alpha > 1$ and*

$$c \le \frac{\sqrt{\pi}\Gamma\left(\frac{\alpha}{2}\right)}{\Gamma\left(\frac{\alpha-1}{2}\right)}.$$

Let us mention that the latter result is no longer true for $\alpha > 2$. It is an open problem to determine the largest constant c that allows (11.4) when $\alpha > 2$.

Proof of Theorem 11.1. The necessity of the condition was proved in [49, Theorem 3.3], so it is enough to prove that if $1 < \alpha \le 2$ and $c = \sqrt{\pi}\Gamma\left(\frac{\alpha}{2}\right)/\Gamma\left(\frac{\alpha-1}{2}\right)$, then there are polynomials with the property (11.4).

Let us consider the energy problem on $\Sigma = [-1,1]$ with weight $w(x) = \exp(c|x|^\alpha)$ (note the positive sign in the exponent which makes this weight essentially different from the Freud weights). First we show that the corresponding extremal measure is given by the density function

$$(11.5) \qquad v(x) = |x|^{\alpha-1}\frac{\alpha}{\pi}\left(\int_0^{|x|} \frac{u^{2-\alpha}}{(1-u^2)^{3/2}} du + d_\alpha\right),$$

where the constant d_α is

$$d_\alpha = \int_0^1 \frac{1 - u^{2-\alpha}}{(1 - u^2)^{3/2}} du.$$

To get this form let us consider the function

$$f(x) = \alpha \frac{1}{\pi\sqrt{1 - x^2}} - (\alpha - 1)\frac{\alpha}{\pi} \int_{|x|}^1 \frac{u^{\alpha-1}}{\sqrt{u^2 - x^2}} du$$

built up from the Chebishev and Ullman distributions (see (3.4)), and recall that the Ullman distribution corresponds to the energy problem with respect to the weight $\exp(-\gamma_\alpha|x|^\alpha)$ (note the negative sign, which is not the case with the energy problem we are discussing now). This function has total integral 1 over $[-1, 1]$, and by (3.6) its logarithm for $x \in [-1, 1]$ is of the form const $+ (\alpha - 1)\gamma_\alpha|x|^\alpha$, where

$$\gamma_\alpha = \frac{\Gamma\left(\frac{\alpha}{2}\right)\Gamma\left(\frac{1}{2}\right)}{2\Gamma\left(\frac{\alpha+1}{2}\right)}.$$

But using that $\Gamma(t + 1) = t\Gamma(t)$ and $\Gamma(1/2) = \sqrt{\pi}$ we easily obtain that

$$(\alpha - 1)\gamma_\alpha = (\alpha - 1)\frac{\Gamma\left(\frac{\alpha}{2}\right)\Gamma\left(\frac{1}{2}\right)}{2\Gamma\left(\frac{\alpha+1}{2}\right)} = \frac{\sqrt{\pi}\Gamma\left(\frac{\alpha}{2}\right)}{\Gamma\left(\frac{\alpha-1}{2}\right)} = c,$$

and so the potential is of the form const $+ c|x|^\alpha$. Thus, if we can show that the function f is nonnegative, then we can invoke Lemma 5.1 to conclude that f is nothing else than the density of the equilibrium measure in question (actually, the same conclusion can be derived from the principle of domination without referring to the nonnegativity of f, but we shall need the following consideration anyway).

Clearly, it is enough to consider positive values of x. If we write the integral in f in the form

$$x^{\alpha-1} \int_1^{1/x} \frac{u^{\alpha-1}}{\sqrt{u^2 - 1}} du$$

and integrate by parts then it follows that f satisfies the differential equation

$$f'(x) = \frac{\alpha}{\pi}\frac{x}{(1 - x^2)^{3/2}} + \frac{\alpha - 1}{x}f(x)$$

with initial condition $f(0) = 0$. We can solve this linear equation and get with some constant d that

$$f(x) = x^{\alpha-1}\frac{\alpha}{\pi}\left(\int_0^x \frac{u^{2-\alpha}}{(1 - u^2)^{3/2}} du + d\right).$$

The value of d follows from the condition that f has integral 1 over $[-1, 1]$. This means that we must have

$$\frac{d}{\pi} + \int_0^1 x^{\alpha-1}\frac{\alpha}{\pi}\int_0^x \frac{u^{2-\alpha}}{(1 - u^2)^{3/2}}\, du dx = \frac{1}{2},$$

which easily yields the value d_α for d. Since d_α is nonnegative, the nonnegativity of f follows from the preceding expression for $f(x)$, and the same expression verifies (11.5).

When $\alpha = 2$ then $d_\alpha = 0$ and (11.5) takes the form

$$v(x) = \frac{2}{\pi} \frac{x^2}{(1 - x^2)^{1/2}},$$

while for $1 < \alpha < 2$ the constant $d_\alpha > 0$ and in this case the density v has order $\sim |x|^{\alpha-1}$ as x approaches 0. Thus, if $\delta = \alpha - 1$ if $1 < \alpha < 2$ and $\delta = 2$ if $\alpha = 2$ then we can conclude that the density v of the extremal measure satisfies $v(t) \sim |t|^\delta$ as $t \to 0$ and $v(t) \sim (1 - t^2)^{-1/2}$ as $t \to \pm 1$, and otherwise v is continuous and positive. This is all we need of v.

Let now $\mu(t) = v(t)dt$ be the extremal measure. By Theorem A from the introduction we have $U^\mu(x) = c|x|^\alpha + F_w$ for every $x \in [-1, 1]$. If we can construct polynomials

$$R_n(x) = \prod_{j=0}^{n-1} (x - \xi_j)$$

such that

(11.6) $-\log|R_n(x)| - nU^\mu(x) \geq C$

for all $x \in [-1, 1]$, and

(11.7) $-\log|R_n(0)| - nU^\mu(0) \leq C,$

then $P_n(x) = R_n(x)/R_n(0)$ will satisfy (11.4). This is where the discretization technique of Section 2 enters the picture.

Let n be an even number (when n is odd, use $n - 1$ in place of n below). Let us divide $[-1, 1]$ by the points $-1 = t_0 < t_1 < \ldots < t_n = 1$ into n intervals I_j, $j = 0, 1, \ldots, n - 1$ with $\mu(I_j) = 1/n$, and let ξ_j be the weight point of the restriction of μ to I_j. Set

$$R_n(t) = \prod_{j=0}^{n-1} (t - \xi_j).$$

We claim that these satisfy (11.6) and (11.7).

We write

(11.8) $-\log|R_n(x)| - nU^\mu(x) = \sum_{j=0}^{n-1} n \int_{I_j} \log\left|\frac{x - t}{x - \xi_j}\right| v(t)dt =: \sum_{j=0}^{n-1} L_j(x).$

The proof of (11.7) is very simple: since $\xi_{n/2-1} < t_{n/2} = 0 < \xi_{n/2}$, and the function $\log|0 - t|$ is concave on every I_j, we get that every term $L_j(x)$ in (11.8) is at most 0. This proves (11.7).

It is left to prove (11.6). Let $x \in I_{j_0}$. The individual terms in (11.8) are clearly bounded from below when $j = j_0, j_0 \pm 1$. For other j's the integrands are bounded in absolute value by an absolute constant independent of n, x and $j \neq j_0, j_0 \pm 1$, hence the integrals themselves are also uniformly bounded, for the integral of v on each I_j equals $1/n$.

As we have done in Section 2, we write for $x \in I_{j_0}$, and $j \neq j_0, j_0 \pm 1$ the integrand in $L_j(x)$ as

$$\log\left|1 + \frac{\xi_j - t}{x - \xi_j}\right| = \frac{\xi_j - t}{x - \xi_j} + O\left(\left|\frac{\xi_j - t}{x - \xi_j}\right|^2\right),$$

which holds because

$$\frac{\xi_j - t}{x - \xi_j} \geq -q > -1, \qquad t \in I_j,$$

with an absolute constant $0 < q < 1$. Thus, we have

$$(11.9) \qquad L_j(x) = n \int_{I_j} O\left(\left|\frac{\xi_j - t}{x - \xi_j}\right|^2\right) v(t)dt$$

$$= O\left(\frac{|I_j|^2}{(\xi_j - \xi_{j_0})^2}\right),$$

because the integrals

$$\int_{I_j} \frac{\xi_j - t}{x - \xi_j} v(t)dt$$

vanish by the choice of the points ξ_j.

We have to distinguish two cases according as x is closer to one of the endpoints or it is closer to 0.

Case I. x is close to an endpoint. Let us suppose for example that $x \in [-1, -1/2]$. We have to estimate

$$S_1(x) := \sum_{j=0}^{j_0-2} |L_j(x)|$$

and

$$S_2(x) := \sum_{j_0+2}^{n-1} |L_j(x)|.$$

We shall only do the first one, the second one being similar (in view of (11.9) the part of S_2 corresponding to the indices for which $\xi_j > -1/4$ is less than

$$(11.10) \qquad C\sum_j |I_j|^2 \leq C\sum_j |I_j| \leq C.)$$

For $j \leq j_0 - 2$

$$1 + \xi_j \sim \left(\frac{j+1}{n}\right)^2, \qquad |I_j| \sim \frac{j+1}{n^2}, \qquad \xi_{j_0} - \xi_j \sim \left(\frac{j_0}{n}\right)^2 - \left(\frac{j}{n}\right)^2,$$

hence

$$S_1(x) \leq C \sum_{j=0}^{j_0-2} \frac{\left(\frac{i+1}{n^2}\right)^2}{\left(\left(\frac{j_0}{n}\right)^2 - \left(\frac{i}{n}\right)^2\right)^2} = \sum_{j=0}^{j_0/2} + \sum_{j=j_0/2}^{j_0-2} =: K_1 + K_2.$$

Here

$$K_1 \leq C \sum_{j=0}^{j_0/2} \frac{(j+1)^2}{j_0^4} = O(1)$$

and

$$K_2 \leq C \sum_{j=j_0/2}^{j_0-2} \frac{j_0^2}{((j-j_0)j_0)^2} = O(1),$$

and these verify that

$$S_1(x) = O(1).$$

Case II. x is close to 0. Now suppose that $x \in [-1/2, 1/2]$, say $x \in [0, 1/2]$. Let $\xi_j^* = \xi_{n/2+j-1}$, $I_j^* = I_{n/2+j-1}$, $L_j^* = L_{n/2+j-1}$. Then $\xi_{-j+1}^* = -\xi_j^*$, $I_{-j+1}^* = -I_j^*$, and for $j > 0$

$$\xi_j^* \sim \left(\frac{j}{n}\right)^{1/(\delta+1)}, \qquad |I_j^*| \sim \frac{1}{n^{1/(\delta+1)}j^{\delta/(\delta+1)}},$$

where δ is the number chosen above, i.e. $\delta = \alpha - 1$ if $1 < \alpha < 2$ and $\delta = 2$ if $\alpha = 2$.

If $x \in I_{j_0}^*$, $j_0 > 0$ then we have to estimate

$$S_1(x) := \sum_{j=1}^{j_0-2} |L_j^*(x)|,$$

$$S_2(x) := \sum_{j=-n/4}^{0} |L_j^*(x)|$$

and

$$S_3(x) := \sum_{j=j_0+2}^{n/4} |L_j^*(x)|,$$

because the contribution of the rest (like that of $\sum_{-n/2}^{-n/4}$) is easily seen to be bounded (use the argument of (11.10)).

We shall estimate $S_1(x) + S_2(x)$, the sum $S_3(x)$ can be similarly handled. Now (11.9) yields

$$
S_1(x) \le C \sum_{j=1}^{j_0-2} \frac{\left(\frac{1}{n^{1/(\delta+1)} j^{\delta/(\delta+1)}}\right)^2}{\left(\left(\frac{j_0}{n}\right)^{1/(\delta+1)} - \left(\frac{j}{n}\right)^{1/(\delta+1)}\right)^2} = \sum_{j=1}^{j_0/2} + \sum_{j=j_0/2}^{j_0-2}
$$

$$
\le C \sum_{j=1}^{j_0/2} \frac{j^{-2\delta/(\delta+1)}}{j_0^{2/(\delta+1)}} + \sum_{j=j_0/2}^{j_0-2} \frac{j^{-2\delta/(\delta+1)}}{((j_0-j)j_0^{-\delta/(\delta+1)})^2} = O(1)
$$

and

$$
S_2(x) \le C \sum_{j=-n/4}^{0} \frac{\left(\frac{1}{n^{1/(\delta+1)}(|j|+1)^{\delta/(\delta+1)}}\right)^2}{\left(\left(\frac{j_0}{n}\right)^{1/(\delta+1)} + \left(\frac{|j|}{n}\right)^{1/(\delta+1)}\right)^2} = \sum_{j<-j_0} + \sum_{j=-j_0}^{0}
$$

$$
\le C \sum_{j<-j_0} \frac{1}{j^2} + \sum_{j=-j_0}^{0} \frac{(|j|+1)^{-2\delta/(\delta+1)}}{j_0^{2/(\delta+1)}} = O(1)
$$

12   Approximation by $W(a_n x)P_n(x)$

Let $W(x) = \exp(-Q(x))$ be a weight function on the real line. In this section we shall consider the problem of approximating functions by weighted polynomials of the form $W(a_n x)P_n(x)$ with some appropriately chosen normalization constants a_n. This type of approximation has been the key to many recent results concerning the orthogonal polynomials with respect to $W(x)$, and the monograph [28] contains the basic results concerning it. If $W(x) = \exp(-c|x|^\alpha)$ is a Freud weight and $a_n = n^{1/\alpha}$, then $W(a_n x)P_n(x) = W(x)^n P_n(x)$, hence in this case our problem is just the one considered in the rest of the present work. In general, we shall set $w_n(x) = W(a_n x)^{1/n}$, so that $W(a_n x)P_n(x) = w_n(x)^n P_n(x)$, and the approximation problem in question is the one considered in Section 8 with varying weights $\{w_n\}$.

On $W(x) = \exp(-Q(x))$ we shall always assume that Q is even, the derivative of $Q(x)$ exists in $(0, \infty)$ and $xQ'(x) \nearrow \infty$ as $x \to \infty$. As normalizing constants we shall use the so called Mhaskar–Rahmanov–Saff numbers a_n, that are defined for sufficiently large n as the solution of the equation

$$
n = \frac{2}{\pi} \int_0^1 \frac{a_n t Q'(a_n t)}{\sqrt{1-t^2}} \, dt.
$$

It is known ([34]) that the supremum norm of weighted polynomials of the form $W(x)R_n(x)$, $\deg R_n \le n$ lives on $[-a_n, a_n]$. So by contraction the norm of weighted polynomials $W(a_n x)P_n(x)$ lives on $[-1,1]$, hence if a function f is the uniform limit of such polynomials then it must vanish outside $(-1,1)$ (see also the end of the proof of Theorem 12.1). Now the problem we face is under what condition is it true that every continuous function f that vanishes outside $(-1,1)$ is the uniform limit of weighted polynomials $W(a_n x)P_n(x)$ (on **R**, or what amounts the same on some interval $[-1-\theta, 1+\theta]$, $\theta > 0$). If the answer to this problem is yes, then we say that the approximation problem for W of type II (as opposed to the problem we have considered previously) is solvable. We shall see that special role is played by the point zero, so we start with a result in which the approximation is guaranteed with a restriction at the origin.

Theorem 12.1 *Let $xQ'(x) \nearrow \infty$ as $x \to \infty$, and suppose that there are $C > 1$ and $\epsilon > 0$ such that*

$$CQ'(Cx) \geq 2Q'(x)$$

and

$$Q'((1+t)x) \leq Q'(x)(1 + Ct^\epsilon)$$

are satisfied for $x \geq x_0$ and $0 < t < 1$. Then every continuous f that vanishes outside $(-1,1)$ and at the origin is the uniform limit of weighted polynomials of the form $W(a_n x)P_n(x)$, $\deg(P_n) \leq n$.

This settles the approximation problem under a rather weak ($C^{1+\epsilon}$-type) smoothness assumption on Q provided we assume $f(0) = 0$ for the function f that we want to approximate (compare this with the results of [28] — especially [28, Theorem 12.2] — where similar results are proved under more restrictive conditions). What happens if we want to approximate a function f that does not vanish at the origin is rather interesting (note that it is enough to approximate *some function* of this type, then every other one can be approximated in view of the preceding result). For completeness we shall briefly discuss the situation along the arguments of [30].

The proof of Theorem [30, Theorem 1] shows that this type of approximation is closely connected with the problem of S. N. Bernstein if for every continuous g with the property

$$W(x)g(x) \to 0 \qquad \text{as} \quad x \to \infty$$

there are polynomials S_n such that

$$\|W(g - S_n)\|_{\mathbf{R}} \to 0 \qquad \text{as} \quad n \to \infty.$$

It is known (see e.g. [1, Theorems 3,5]) that in our case (i.e. when $xQ'(x)$ increases to infinity) the necessary and sufficient condition for a positive answer to Bernstein's problem is

$$\int_{-\infty}^{\infty} \frac{\log W(t)}{1 + t^2}\, dt = \infty.$$

Theorem 12.2 *If in the case*

$$\int_{-\infty}^{\infty} \frac{\log W(t)}{1+t^2}\, dt < \infty$$

there are polynomials P_n of degree at most n such that

$$\|W(a_n x)P_n(x) - f(x)\|_{\mathbf{R}} \to 0 \qquad \text{as} \quad n \to \infty$$

for some f that does not vanish at the origin, then W^{-1} must be an entire function.

On the other hand, we have

Theorem 12.3 *Let*

$$\int_{-\infty}^{\infty} \frac{\log W(t)}{1+t^2}\, dt = \infty,$$

and suppose that $Q(x)$ is twice continuously differentiable for large x, and the function

$$T(x) = \frac{(xQ'(x))'}{Q'(x)},$$

lies in between two positive constants:

(12.1) $$0 < A \leq T(x) \leq B$$

for $x \geq x_0$ and is of slow variation in the sense that

$$\lim_{x \to \infty} \frac{T(\lambda x)}{T(x)} = 1$$

for all $\lambda > 0$. Then the approximation problem of type II is solvable for W.

These results can be applied for example to the Freud weights $W(x) = \exp(-c|x|^\alpha)$, $c, \alpha > 0$, in which case

$$a_n = \gamma_\alpha^{1/\alpha} c^{-1/\alpha} n^{1/\alpha}, \qquad \gamma_\alpha := \Gamma(\frac{\alpha}{2})\Gamma(\frac{1}{2})/(2\Gamma(\frac{\alpha}{2}+\frac{1}{2}))$$

is the number from Section 1. We can conclude that for all $\alpha > 0$ every f that vanishes outside $(-1, 1)$ and at the origin is the uniform limit of weighted polynomials $W(a_n x)P_n(x)$. When $f(0) \neq 0$, then this is the case if and only if $\alpha \geq 1$.

Proof of Theorem 12.1. We set

$$Q_n(x) = Q(a_n x)/n, \quad \text{and} \quad w_n(x) = \exp(-Q_n(x)).$$

Our conditions immediately imply that $Q'_n(x) \sim Q'_n(y)$ if $x \sim y$, $x, y \to \infty$, and that for any fixed $x_0 > 0$ we have $a_n x_0 Q'(a_n x_0) \sim n$, which can be translated as $Q_n'(x_0) \sim 1$ (here $A \sim B$ denotes that the ratio A/B stays away from zero and infinity in the range considered). Taking into account also that $xQ_n'(x)$ is nondecreasing on $(0,1)$, we can easily conclude from the assumptions of the theorem that the Q_n's uniformly belong to $C^{1+\epsilon}$ on every compact subinterval of $(0,1)$, furthermore there are constants $0 < c < d < 1$ and $\eta > 0$ such that $dQ_n'(d) \geq cQ_n'(c) + \eta$. Hence, Corollary 8.4 can be applied and we obtain the statement of the theorem, at least for concluding uniform convergence on $[-1,1]$.

However, we know that in Corollary 8.4 the convergence is actually true on a larger set $[-1-\theta, 1+\theta]$, $\theta > 0$, and outside this interval weighted polynomials $W(a_n x)P_n(x)$ that are bounded on $[-1,1]$ automatically tend to zero under our conditions. This can be seen as follows.

For any equilibrium measure $\mu = \mu_w$ with $w(x) = \exp(-Q(x))$, $xQ'(x) \nearrow$, $S_w = \mathrm{supp}(\mu_w)$ we have that

$$x\left(U^\nu(x) + Q(x)\right)' = -\int_{-1}^1 \frac{x}{x-t}d\mu(t) + xQ'(x)$$

increases on $[1,\infty)$, furthermore this expression is nonnegative around $x = 1$ (see Theorem A,(d) and (f) in Section 1). Since for all $t \in [-1,1]$ and $x \geq 1+\theta/2$

$$\frac{x}{x-t} \geq \frac{1+\theta/4}{1+\theta/4-t} + \frac{\theta}{16},$$

we can conclude (see also Theorem A,(f)) that for $x \geq 1 + \theta$ the sum $U^\mu(x) + Q(x) - F_w$ is at least as large as $\theta^2/64$. Now we can make use of Theorem B from the introduction according to which for such x

$$|w^n(x)P_n(x)| \leq M \exp\left(n(-Q(x) - U^\mu(x) + F_w)\right) \leq \exp(-n\theta^2/64).$$

Apply now this to $w(x) = W(a_n x)^{1/n}$.   ∎

Proof of Theorem 12.2. Suppose that $f(0) \neq 0$, say $f(0) = 1$, and there are polynomials P_n with $W(a_n x)P_n(x) - f(x)$ uniformly tending to zero on $\mathbf{R}$. Set $f_n(x) = f(x/a_n)$ and $R_n(x) = P_n(x/a_n)$. Then we get that

$$\|f_n - WR_n\|_{\mathbf{R}} \to 0$$

as $n \to \infty$, and hence

$$\|W(W^{-1}f_n - R_n)\|_{\mathbf{R}} \to 0,$$

which implies

$$\|W^2(W^{-1}f_n - R_n)\|_{\mathbf{R}} \to 0.$$

We also have

$$\|W^2 W^{-1}\|_{\mathbf{R} \setminus [-\gamma, \gamma]} \to 0 \qquad \text{as } \gamma \to \infty,$$

furthermore the functions f_n uniformly tend to $f(0) = 1$ on compact subsets of **R** (note that $a_n \to \infty$ as $n \to \infty$). These imply

$$\|W^2(W^{-1} - R_n)\|_{\mathbf{R}} \to 0, \quad \text{as } n \to \infty.$$

But then the polynomials $\{R_n\}$ are uniformly bounded on every compact subset of the complex plane (see e.g. [1, Theorems 5,7]), hence we can select a subsequence from them that converges to an entire function on the whole plane. But $R_n(x) \to W(x)^{-1}$ for every real x, so W^{-1} has to be the entire function in question.

                                                        ∎

Proof of Theorem 12.3. First of all let us mention that in view of Theorem 12.1 it is enough to show the following: for every $\epsilon > 0$ there is a continuous function χ (that may also depend on ϵ) such that $\chi(0) = 1$, and for all sufficiently large n there are polynomials P_n of degree at most n such that $|W(x)P_n(x) - \chi(x/a_n)| \leq \epsilon$ for all $x \in \mathbf{R}$. In fact, suppose this is true and we want to approximate an f which vanishes outside $(-1, 1)$. Then, in view of Theorem 12.1, for a given $\epsilon > 0$ we can approximate $f(x) - f(0)\chi(x)$ by a $W(a_n x)R_n(x)$ uniformly on **R** with error smaller than ϵ. The sum $P_n(a_n x) + R_n(x)$ multiplied by $W(a_n x)$ will then be closer to f than 2ϵ.

We shall need a lemma, which is a variant of a result of D. S. Lubinsky [24].

Lemma 12.4 *Under the conditions of Theorem 12.3 there is an even entire function H with nonnegative McLaurent coefficients such that $W(x)H(x) \to 1$ as $x \to \infty$.*

Proof of Lemma 12.4. The lemma can be proven with the method of [24, Theorems 5,6]. We shall only indicate how the construction goes and what changes are necessary in those proof.

As before, let $W(x) = \exp(-Q(x))$, and for $x > 0$ let the number q_x be the solution of the equation $q_x Q'(q_x) = 2x$. Then

$$(12.2) \qquad\qquad\qquad q_n \sim a_n.$$

With the function $T(x)$ from the theorem we set

$$H(x) = \sum_{n=0}^{\infty} \left(\frac{x}{q_n}\right)^{2n} \frac{1}{\sqrt{\pi n T(Q(n))}} e^{Q(q_n)}.$$

This H will satisfy the claim in the lemma. The proof is an adaptation of that of [24, Theorem 5], we are not going into the details.

                                                        ∎

Now let us return to the proof of Theorem 12.3. Let R_m be the m-th partial sum of the McLaurent expansion of the entire function from the preceding lemma. Since the coefficients of H are nonnegative, we have for any x the inequality $0 \leq R_m(x) \leq H(x)$, furthermore, $R_m(x) \to H(x)$ uniformly on compact subsets of $\mathbf{R}$. As for the Mhaskar–Rahmanov–Saff numbers a_n, we can conclude from the assumption (12.1) that there is a positive constant C such that $a_n \leq a_{2n} \leq C a_n$ holds for all large n. This and (12.2) easily imply (see also the computation for q_x in [24]) that given any $\epsilon > 0$ there are numbers r and t such that

$$\frac{|H(x) - R_m(x)|}{H(x)} \leq \epsilon \qquad \text{for } |x| \leq r a_m$$

and

$$\frac{|R_m(x)|}{H(x)} \leq \epsilon \qquad \text{for } |x| \geq t a_m.$$

Now $W(x)H(x) - 1$ tends to zero at infinity, hence by the solution to Bernstein's problem to every $\epsilon > 0$ there exists a polynomial S such that for all $x \in \mathbf{R}$

$$|W(x)(W^{-1}(x) - H(x) - S(x))| \leq \epsilon.$$

This gives that

$$|1 - W(x)(H(x) - S(x))| \leq \epsilon,$$

and so in view of of the preceding inequalities we obtain that

$$(12.3) \qquad |1 - W(x)(R_m(x) - S(x))| \leq (1 + M)\epsilon \qquad \text{for } |x| \leq r a_m,$$

$$(12.4) \qquad W(x)|R_m(x) - S(x)| \leq M\epsilon \qquad \text{for } |x| \geq t a_m,$$

and

$$(12.5) \qquad W(x)|R_m(x) - S(x)| \leq M(M + 2)$$

otherwise, where M is an upper bound for WH on $\mathbf{R}$.

The assumptions on Q imply (see also the computation for q_x in [24]) that there is also an L such that $2t a_{[n/L]} \leq r a_n$ is also true. Now if we set

$$P_n(x) = \frac{1}{l} \sum_{k=1}^{l} (R_{[nL^{-k}]}(x) - S(x)),$$

then we ge from (12.3)—(12.5) that

$$|W(x)P_n(x) - \chi(|x|/a_n)| \leq (1 + M)\epsilon + \frac{M(M + 2)}{l},$$

where the function $\chi(x)$ is defined on $[0, \infty)$ as follows: $\chi(L^{-k}) = 1 - 1/k$ for $k = 1, 2, \ldots, l - 1$, $\chi(0) = 1$, $\chi(x) = 0$ for $x \geq 1$ and χ is linear otherwise. Hence the polynomials P_n and the function χ satisfy the requirement from the beginning of the present proof if we choose $l > M(M + 2)/\epsilon$.

13   Extremal problems with varying weights

Let $\{w_n\}$ be a sequence of weights on the interval $[-1, 1]$, u a fixed nonnegative function and for a $1 \leq p < \infty$ consider the extremum problem

$$(13.1) \qquad E_{m,p}(w_n^n u) := \inf_{P_m \in \Pi_m} \|P_m w_n^n u\|_{L^p},$$

where again Π_m denotes the set of polynomials $\{x^m + \cdots\}$ with leading coefficient 1. We have already discussed this in Section 3.3 in the particular case when each w_n is the same. From Theorems 10.1 and 10.2 and Bernstein's formula (3.38) we can easily get the following strong asymptotics for $E_{m,p}$.

Theorem 13.1 *Let $1 \leq p < \infty$, $\{w_n\}$ a sequence of weight functions on the interval $[-1, 1]$ such that the corresponding extremal measures μ_{w_n} have support $[-1, 1]$, they are absolutely continuous there and if we write $d\mu_{w_n}(t) = v_n(t) \, dt$, then the functions v_n satisfy the conditions*

$$(13.2) \qquad v_n(t) \leq A(1 - t^2)^\beta, \qquad t \in (0, 1),$$

$$(13.3) \qquad v_n(t) \geq \frac{1}{A}(1 - t^2)^{\beta_0}, \qquad t \in (0, 1)$$

for some constants A, $\beta > -1$ and β_0. Let furthermore u be a positive continuous function. Then for every $k = 0, \pm 1, \ldots$

$$(13.4) \qquad E_{n+k,p}(w_n^n u) = (1 + o(1))\sigma_p 2^{-n-k+1-1/p} G[w_n]^n G[u]$$

as $n \to \infty$. Actually, the relation (13.4) is uniform in $k \geq -K$ for every fixed K.

The result is also true for even weights $\{w_n\}$ satisfying on $[0, 1]$ the conditions of Theorem 10.3.

Recall that

$$G[V] = \exp\left(\frac{1}{\pi} \int_{-1}^1 \frac{\log V(x)}{\sqrt{1 - x^2}} \, dx\right)$$

are the geometric means from (3.13) and

$$\sigma_p = \left(\Gamma(1/2)\Gamma((p+1)/2)/\Gamma(p/2+1)\right)^{1/p}.$$

If we integrate the equality in Theorem A,(f) with respect to the equilibrium measure of the interval $[-1, 1]$ and use Funibi's theorem as well as (1.7)–(1.8), then we get that

$$G[w_n] = \exp(\log 2 - F_{w_n}),$$

hence (13.4) can be written in the form

$$E_{n+k,p}(w_n^n u) = (1 + o(1))\sigma_p 2^{-k+1-1/p} G[u] e^{-n F_{w_n}}.$$

The result holds for somewhat more general u's, but the exact conditions on u are not clear. Below we shall prove a more general result for $p = 2$. This case corresponds to orthogonal polynomials and appears in several situations.

Proof. . The proof is very similar to the argument of Section 3. First let us assume that $k \geq 0$. The assumptions imply that Theorems 10.1 and 10.2 can be applied on the interval $(-1, 1)$ rather than on $(0, 1)$. We set $\gamma = (1 - 1/p)/2$. Then by Theorem 10.1 there is a sequence $\{H_n\}$ of polynomials of corresponding degree $n = 1, 2, \ldots$ such that if

$$h_n(x) = w_n^n(x)u(x)|H_n(x)|(1 - x^2)^{(-1+1/p)/2},$$

then $h_n(x) \geq 1$ and

$$\lim_{n \to \infty} G[h_n] = 1.$$

Thus, by Bernstein's formula (3.38)

$$\liminf_{n \to \infty} E_{n+k,p}(w_n^n u) \Big/ \left(\sigma_p 2^{-n-k+1-1/p} G[w_n]^n G[u] \right)$$

$$\geq \liminf_{n \to \infty} E_{n+k,p}(\varphi^{1/2-1/2p}/|H_n|) \Big/ \left(\sigma_p 2^{-n-k+1-1/p} G[w_n]^n G[u] \right)$$

$$= \liminf_{n \to \infty} G[\varphi^{1/2-1/2p}/|H_n|] \Big/ (G[w_n]^n G[u]),$$

and if we consider that here the fraction on the right hand side is $1/G[h_n]$, it follows that

$$\liminf_{n \to \infty} E_{n+k,p}(w_n^n u) \Big/ \left(\sigma_p 2^{-n-k+1-1/p} G[w_n]^n G[u] \right) \geq 1.$$

The proof of the upper estimate is completely symmetric if we use Theorem 10.2 and the polynomials $R_{2n}(x) = |H_{n-1}(x)|^2(1 - x^2)$ there which are positive in $(-1, 1)$ with simple zeros at ± 1 (c.f. (3.38)).

Since the degree of H_n in the proof can be smaller than n, say $n - i_n$ with $i_n \to \infty$, it also follows that the preceding argument actually holds for all k because eventually we shall have $k \geq -i_n$.

The uniformity of the convergence in $k \geq -K$ also follows from the proof.

Finally, the last statement follows from Corollary 10.4 since the proof used only the existence of the polynomials guaranteed by this corollary (in Corollary 10.4 the degree of the polynomials H_n can again be $n - i_n$ where $i_n \to \infty$). ∎

We have already mentioned that the quantity $E_{n,2}(W)$ gives the reciprocal of the leading coefficient of the n-th orthonormal polynomial with respect to the weight function W^2. Thus, the case $p = 2$ is of special interest and is connected with many other problems in mathematics. Our next aim is to extend Theorem 13.1 in this case to more general weight functions u. This extension is connected to multipoint Padé approximation that we shall briefly discuss in Section 16.

Theorem 13.2 *Let $p = 2$. Then with the assumptions of Theorem 13.1 the asymptotic relation (13.4) holds for every measurable u that is positive almost everywhere on $[-1, 1]$.*

We also add that in this case $\sigma_2 = \sqrt{\pi/2}$, so (13.4) has the form

$$E_{n+k,2}(w_n^n u) = (1 + o(1))\sqrt{\pi}2^{-n-k}G[w_n]^n G[u].$$

Let us also mention that if u does not satisfy the the so called Szegő condition

(13.5)
$$\int_{-1}^{1} \frac{\log u(t)}{\sqrt{1-t^2}}\, dt > -\infty,$$

then formula (13.4) gives only an upper estimate. Thus, in finer asymptotic problems we shall assume (13.5) exactly as is done in the classical case.

Proof.   We can assume that u satisfies the Szegő condition (13.5). In fact, then the general case when this is not so follows by adding to u a positive δ and then letting δ tend to zero.

The case when u is continuous and positive follows from Theorem 13.1. The general case will be deduced with the aid of a theorem of G. Lopez [19]. Lopez proved the following: let

$$V_{2n}^* = \prod_{j=1}^{2n-j_n} (x - z_{n,j})$$

be polynomials of degree $2n - j_n$ with $j_n \to \infty$ which are positive on $(-1, 1)$, and let μ be a measure which has positive (Radon–Nikodym) derivative almost everywehere in $(-1, 1)$. We set $d\mu_n = (V_{2n}^*)^{-1}d\mu$, and assume that μ_n is a finite measure on $(-1, 1)$ for every n. Then we can form the orthonormal polynomials $p_{n,k}^*$ with respect to μ_n:

$$\int_{-1}^{1} p_{n,k}^* p_{n,m}^* \, d\mu_n = \delta_{k,m}.$$

Assume further, that if $\Phi(z) = z + \sqrt{z^2 - 1}$ is the conformal map that carries the complement of $[-1, 1]$ into the unit circle, then for the zeros of the V_{2n}^*'s the condition

$$\lim_{n \to \infty} \sum_{j=1}^{2n-j_n} (1 - |\Phi(z_{n,j})|^{-1}) = \infty$$

is satisfied. Under these conditions we have ([19, Theorem 9]) for $n \to \infty$

(13.6)
$$\lim_{n \to \infty} \int_{-1}^{1} f(t)\frac{(p_{n,n+k}^*)^2(t)}{V_{2n}^*(t)}\, d\mu(t) = \frac{1}{\pi}\int_{-1}^{1} f(t)\frac{1}{\sqrt{1-t^2}}\, dt$$

for every bounded f and every fixed $k = 0, \pm 1, \ldots.$ In other words, the measures $(p_{n,n+k}^*)^2 d\mu_n$ converge weakly to the arcsine distribution in the stronger sense indicated.

Let $\gamma_{n,k}^*$ be the leading coefficient of $p_{n,k}^*$, i.e.

$$p_{n,k}^*(x) = \gamma_{n,k}^* x^k + \cdots.$$

As we have already mentioned, it is immediate from the orthogonality of the polynomials $p_{n,k}^*$ that the extremal polynomial in the minimum problem (13.1) (with $p = 2$) for the measure $(V_{2n}^*)^{-1} d\mu$ is $p_{n,k}^*/\gamma_{n,k}^*$, i.e.

$$\inf_{P_n \in \Pi_n} \int |P_n|^2 \frac{1}{V_{2n}^*} \, d\mu = \int \left(\frac{p_{n,k}^*}{\gamma_{n,k}^*}\right)^2 \frac{1}{V_{2n}^*} \, d\mu.$$

It also follows from the considerations of [19] and [21] that (see also [19, (43)])

(13.7) $\gamma_{n,n+k}^* = (1 + o(1))\sqrt{\pi} 2^{n+k} G[V_{2n}^*]^{n/2} G[\mu']^{-1/2}.$

In fact, the corresponding result for the unit circle was proved in [21] under the assumptions that the zeros of certain polynomials (the analogues of V_{2n}) are real, but the proof works just as well for the general case if one uses the results of [19]. From here the transfer of the result to the real line to obtain (13.7) is just the standard technique (c.f. [33] or [19] and use also [19, Theorem 3]).

After these preliminaries we turn to the proof of Theorem 13.2. By Theorem 10.2 there is a sequence $\{H_{n-1}\}$ of polynomials such that if

$$h_n(x) = w_n^n(x)|H_{n-1}(x)\sqrt{1-x^2}|(1-x^2)^{-1/4},$$

then $h_n(x) \leq 1$ and

$$\lim_{n \to \infty} G[h_n] = 1.$$

Actually, as we have remarked after the lemma, the degree of H_{n-1} can be $n - i_n$ with $i_n \to \infty$. Furthermore, the construction of Lemma 9.1 shows that at least half of the zeros of the H_n are of distance $L_n cn^{-1}$ from the interval $[-1, 1]$, where $L_n \to \infty$ as $n \to \infty$. Thus, if we set

$$V_{2n}^*(x) = |H_{n-1}(x)|^2(1 - x^2)$$

and $d\mu(t) = (1 - t^2)^{1/2} u^2(t) \, dt$, then all the properties of V_{2n}^* mentioned above are satisfied (note that if $\text{dist}(z, [-1, 1]) \geq cL_n/n$, then

$$1 - |\Phi(z)|^{-1} \geq c_1 \min(1, L_n/n)),$$

hence for the corresponding orthogonal polynomials $p_{n,k}^*$ we have (13.7). But $h_n \leq 1$ implies

$$E_{n+k,2}(w_n^n u) \leq E_{n+k,2}\left(\frac{u\varphi^{1/4}}{(V_{2n}^*)^{1/2}}\right),$$

and so it follows from the formula

(13.8) $$E_{n+k,2}\left(\frac{u\varphi^{1/4}}{(V_{2n}^*)^{1/2}}\right) = \frac{1}{\gamma_{n,n+k}^*},$$

the asymptotic relation (13.7) and

$$G[w_n]^n G[u] \Big/ G[\varphi^{1/4} u/(V_{2n}^*)^{1/2}] = G[h_n] = 1 + o(1),$$

that

$$\limsup_{n \to \infty} E_{n+k,2}(w_n^n u) \Big/ \left(\sigma_2 2^{-n-k+1/2} G[w_n]^n G[u]\right) \leq 1$$

The proof of the lower estimate is similar if we use Theorem 10.1 instead Theorem 10.2.

This completes the proof. ∎

14   Asymptotic properties of orthogonal polynomials with varying weights

Let again $\{w_n\}$ be a sequence of weights on $[-1, 1]$ as before, and u a measurable function satisfying the Szegő condition (13.5). We denote by

$$p_{n,k}(x) = \gamma_{n,k}(x) x^k + \cdots, \qquad \gamma_{n,k} > 0,$$

the k-th orthonormal polynomial with respect to $w_n^{2n} u^2$:

$$\int_{-1}^{1} p_{n,k} p_{n,m} w_n^{2n} u^2 = \delta_{k,m}.$$

Note the square in the weight. For the monic polynomials

$$q_{n,k} := \frac{1}{\gamma_{n,k}} p_{n,k}$$

we have

(14.1)
$$\int q_{n,k} q_{n,k} w_n^{2n} u^2 = E_{k,2}^2(w^n u),$$

where $E_{k,2}$ is the extremal quantity discussed in Section 13. This means that the monic orthogonal polynomials are the optimal ones in the extremal L^2 problem of Section 13 for the case $p = 2$, and we also have

(14.2)
$$\gamma_{n,k} = 1/E_{k,2}(w^n u).$$

In this section we discuss asymptotics on $p_{n,k}$. For fixed weights ($w_n \equiv 1$) all these are classical, and some of the results below were proved by G. Lopez [19, 21] for varying case when the w_n^{2n}'s are reciprocals of polynomials.

For simpler notation let W_n be the weight $w_n^n u$.

With some positive $\eta > 0$ we choose a strictly positive continuous function u^* that coincides with u outside a set E_η of measure smaller than η. By Theorem

10.2 (see also the remark after it) there is a sequence $\{H_{n-1}\}$ of polynomials of degree $n - i_n$, $i_n \to \infty$, such that that they do not vanish on $(-1,1)$ and if

$$h_n^*(x) = w_n^n(x)u^*(x)|H_{n-1}(x)\sqrt{1-x^2}|(1-x^2)^{-1/4},$$

then $h_n^*(x) \leq 1$ and

(14.3) $$\lim_{n \to \infty} G[h_n^*] = 1.$$

We also set

(14.4) $$\chi(x) = \min\{u^*(x)/u(x), 1\}$$

which again belongs to the Szegő class. By chosing η sufficiently small (and u^* appropriately) the geometric mean of u^* can be as close as we wish to that of u, hence the geometric means of χW_n and W_n will also be close. Now with

(14.5) $$W_n^* = \frac{\varphi^{1/4}}{|H_{n-1}|\varphi^{1/2}}, \qquad \varphi(x) = 1 - x^2$$

we have

$$\chi(x)W_n(x)/W_n^*(x) \leq h_n^*(x) \leq 1, \qquad x \in [-1,1]$$

and the geometric means of the three weights W_n, χW_n and W_n^* can differ by as small amount as we wish if we choose η and u^* appropriately and n is sufficiently large. These properties and Theorem 13.2 on the asymptotic behavior of the leading coefficient of orthogonal polynomials with respect to varying weights easily imply the following: for every ϵ we have polynomials H_n of degree $n - i_n$ with $i_n \to \infty$ such that for some set E_ϵ of measure at most ϵ we have the relations

(14.6) $$W_n(x)/W_n^*(x) = 1 + o(1), \qquad x \notin E_\epsilon,$$

and for every fixed $k = 0, \pm 1, \cdots$ and large n

(14.7) $$\frac{E_{n+k,2}(\chi W_n)}{2}\left(\frac{1}{E_{n+k,2}(W_n)} + \frac{1}{E_{n+k,2}(W_n^*)}\right) \geq 1 - \epsilon,$$

(14.8) $$E_{n+k,2}^2(\chi W_n) \leq (1+\epsilon)E_{n+k,2}^2(W_n^*)$$

and

(14.9) $$\begin{aligned} E_{n+k,2}^2(\chi W_n) &\leq \int (q_{n,n+k}^*)^2(\chi W_n)^2 \leq \int (q_{n,n+k}^*)^2(W_n^*)^2 \\ &= E_{n+k,2}^2(W_n^*) \leq (1+\epsilon)E_{n+k,2}^2(\chi W_n), \end{aligned}$$

where $q_{n,m}^*$ denotes the m-th monic orthogonal polynomial corresponding to $(W_n^*)^2$.

We have already mentioned that by [19, Theorem 9] the functions

$$(p_{n,n+k}^*W_n^*)^2$$

tend to

$$\omega := \frac{1}{\pi} \frac{1}{\sqrt{1-t^2}}$$

in the sense that for every bounded and measurable f

$$(14.10) \qquad \lim_{n\to\infty} \int f(p_{n,n+k}^* W_n^*)^2 = \int f\omega.$$

We start with a simple observation. We apply the parallelogram law

$$\frac{1}{4}\int (p_{n,n+k} - p_{n,n+k}^*)^2 (\chi W_n)^2 + \int \left(\frac{1}{2}(p_{n,n+k} + p_{n,n+k}^*)\right)^2 (\chi W_n)^2$$

$$= \frac{1}{2}\int (p_{n,n+k})^2 (\chi W_n)^2 + \frac{1}{2}\int (p_{n,n+k}^*)^2 (\chi W_n)^2,$$

and observe that the first term on the right is at most $1/2$ by $\chi \leq 1$, the second one is at most $(1+\epsilon)^2/2$ by (14.9) and (14.8) (see also (14.2)), while the second term on the left is at least as large as $(1-\epsilon)^2$ by (14.7) and (14.2). Therefore, we can conclude that

$$(14.11) \qquad \int (p_{n,n+k} - p_{n,n+k}^*)^2 (\chi W_n)^2 \leq 12\epsilon,$$

in particular,

$$(14.12) \qquad \int_{E_\epsilon^c} (p_{n,n+k} - p_{n,n+k}^*)^2 W_n^2 \leq 12\epsilon,$$

where $E_\epsilon^c := [-1,1] \setminus E_\epsilon$ denotes the complement of E_ϵ in $[-1,1]$ (recall that $\chi(x) = 1$ for all $x \in E_\epsilon^c$).

Let now T be a measurable subset of $[-1,1]$ not intersecting E_ϵ. We can easily get from (14.12), (14.6), (14.10) and Schwarz inequality that

$$\liminf_{n\to\infty} \int_T (p_{n,n+k} W_n)^2 \geq \int_T \omega - 12\epsilon - 2\sqrt{12\epsilon},$$

which gives for $\epsilon \to 0$ that for *every* measurable subset T of $[-1,1]$

$$\liminf_{n\to\infty} \int_T (p_{n,n+k} W_n)^2 \geq \int_T \omega.$$

On applying this to the complement of T in $[-1,1]$, it follows that here the liminf can be replaced by limit (recall, that the corresponding integral over $[-1,1]$ is 1). Since every bounded function can be uniformly approximated by linear combinations of characteristic functions of sets, we finally arrive at

Theorem 14.1 *Let $\{w_n\}$ be a sequence of weight functions on $[-1,1]$ such that the corresponding extremal measures μ_{w_n} have support $[-1,1]$, they are absolutely continuous there and if we write $d\mu_{w_m}(t) = v_n(t)\,dt$, then the functions v_n*

satisfy the conditions (13.2) and (13.3). Let furthermore u satisfy Szegő's condition (13.5). If $p_{n,m}$ are the orthonormal polynomials with respect to $w_n^{2n}u^2$, then for every fixed $k = 0, \pm 1, \dots$ and for every bounded and measurable f

$$\lim_{n \to \infty} \int f(p_{n,n+k} w_n^n u)^2 = \int f\omega.$$

The result is also true for even weights $\{w_n\}$ satisfying on $[0,1]$ the conditions of Theorem 10.3.

For fixed weight, i.e. when $w_n \equiv 1$ this was proved in [33, Theorem 11.1], while, as we have seen and used above, for varying weights when w_n^{2n} is the reciprocal of a polynomial with zeros not too close to $[-1,1]$, by Lopez [19, Theorem 9].

It would be equally easy to extend other results from [33] and [19] to the present case at least under Szegő's condition on u. Probably the results also hold if we only assume that $u(t) > 0$ almost everywhere on $[-1,1]$. Instead of pursuing this direction further we seek stronger asymptotics on the orthogonal polynomials. As one can expect, we shall get strong asymptotics away from $[-1,1]$, and on $[-1,1]$ in L^2 norm.

To this end we shall need the following corollary to Theorem 14.1. With the notations applied before the theorem we have

$$(14.13) \qquad \limsup_{n \to \infty} \int \left(p_{n,n+k} W_n - p_{n,n+k}^* W_n^* \right)^2 \leq \sqrt{\epsilon}$$

as $n \to \infty$. In fact, with the set E_ϵ used before we have

$$\int \left(p_{n,n+k} W_n - p_{n,n+k}^* W_n^* \right)^2 \leq 2 \left(\int_{E_\epsilon} (p_{n,n+k} W_n)^2 + \int_{E_\epsilon} (p_{n,n+k}^* W_n^*)^2 \right)$$

$$+ 2 \int_{E_\epsilon^c} \left(p_{n,n+k} W_n - p_{n,n+k}^* W_n \right)^2 + 2 \int_{E_\epsilon^c} \left(p_{n,n+k}^* (W_n - W_n^*) \right)^2,$$

and by the preceding theorem, (14.10), (14.12) and (14.6) we get the bound

$$\frac{2}{\pi} \int_{E_\epsilon} \omega + 24\epsilon + o(1)$$

for the right hand side, and this is easily seen to be smaller than $\sqrt{\epsilon}$ for large n (and small ϵ).

To formulate our next theorem we need a definition (see [48, (12.2.3)], where an additional factor $1/2$ appears on the right which we put into our functions). If V is a nonnegative function on $[-1,1]$, then let

$$\Gamma_V(x) = \frac{1}{\pi} \int_{-1}^{1} \frac{\log V(\xi) - \log V(x)}{\xi - x} \left(\frac{1 - x^2}{1 - \xi^2} \right)^{1/2} d\xi,$$

where the integral is understood in principal value sense. It is easy to see that $\Gamma_V(\cos\theta)$ coincides for $0 \leq \theta \leq \pi$ with the trigonometric conjugate of $\log V(\cos\theta)$ (see [48, (12.2.3)]).

Theorem 14.2 *Let w_n satisfy the assumptions of Theorem 14.1, and suppose that u satisfies Szegő's condition (13.5). Then for fixed $k = 0, \pm 1, \ldots$ the difference*

$$p_{n,n+k}(x)w_n^n(x)u(x)$$

$$-\sqrt{\frac{2}{\pi}}\frac{1}{\sqrt[4]{1-x^2}}\cos\left(\left(n+k+\frac{1}{2}\right)\arccos x + n\Gamma_{w_n}(x) + \Gamma_u(x) - \frac{\pi}{4}\right)$$

tends to zero in $L^2[-1,1]$.

Proof. We continue to use the functions and notations from before Theorem 14.1. Recall that there we have chosen to an $\epsilon > 0$ a function u^* and polynomials H_{n-1} with properties (14.3)–(14.9).

It follows from [48, Theorem 2.6, (2.6.2)] for $\rho(x) = H_{n-1}^2(x)$ that

$$p_{n,n+k}^*(\cos\theta)W_n^*(\cos\theta)(\sin\theta)^{1/2} = (2/\pi)^{1/2}\Re\left\{e^{i(n+k)\theta}\overline{\exp(i\Gamma_{|H_{n-1}|}(\cos\theta))}\right\}$$

for all large n (so that $n - i_n < n + k$). The last factor on the right is

$$\cos\left((n+k)\theta - \Gamma_{|H_{n-1}|}(\cos\theta)\right),$$

so in view of (14.13) it is enough to prove that the difference of

$$\cos\left((n+k)\theta - \Gamma_{|H_{n-1}|}(\cos\theta)\right)$$

and

$$\cos\left(\left(n+k+\frac{1}{2}\right)\theta + n\Gamma_{w_n}(\cos\theta) + \Gamma_u(\cos\theta) - \frac{\pi}{4}\right)$$

in $L_{2\pi}^2[0,\pi]$ is as small as we wish for small ϵ. Since the conjugate function of $\log|\sin\theta|$ is $\theta - (\pi/2)$ for $\theta \in [0,\pi]$ and $(\pi/2) + \theta$ for $\theta \in [-\pi, 0]$, the last but one expression is actually

$$\cos\left((n+k+\frac{1}{2})\theta + \Gamma_{W_n^*}(\cos\theta) - \frac{\pi}{4}\right),$$

and so it is enough to show that the two functions

$$\Gamma_{W_n^*}(\cos\theta)$$

and

$$\Gamma_{w_n^n u}(\cos\theta)$$

are close on a set of almost full measure on $[0, \pi]$ provided n is sufficiently large. But the difference of these two functions is $\Gamma_{h_n^* u/u^*}(\cos\theta)$, and so recalling that $\Gamma_V(\cos\theta)$ coincides with the trigonometric conjugate of $\log V(\cos\theta)$, it follows from the weak (1,1) property of the operator of trigonometric conjugation that the measure of the set

$$\{\theta \in [0, \pi] \,|\, |\Gamma_{h_n^* u/u^*}(\cos \theta)| \geq \eta\}$$

is at most

$$C\eta^{-1} \int_0^\pi |\log h_n^*(\cos \theta)| \, d\theta$$

$$= C\eta^{-1} \left(\int_0^1 \frac{|\log h_n^*(x)|}{\sqrt{1-x^2}} \, dx + \int_0^1 \frac{|\log u_n(x)/u_n^*(x)|}{\sqrt{1-x^2}} \, dx \right).$$

Since for fixed η the first term on the right hand side tends to zero as $n \to \infty$ (see Theorem 10.1) and the second term is as small as we like by appropriately choosing u^*, the proof is complete.

∎

Finally, we prove a strong asymptotic formula for $p_{n,n+k}$ away from $[-1, 1]$. To this end we introduce the so called Szegő function

$$D_V(z) = \exp\left(\sqrt{z^2 - 1} \frac{1}{2\pi} \int_{-1}^1 \frac{\log V(x)}{z - x} \frac{dx}{\sqrt{1-x^2}} \right)$$

provided $\log V(x)/\sqrt{1-x^2}$ is integrable, i.e. provided V satisfies the Szegő condition (13.5). This form of the Szegő function appears in [22], and can be deduced from the more familiar one corresponding to the unit disk by the standard conformal mapping between $C \setminus [-1, 1]$ and the unit disk. The Szegő function for the unit disk is (see e.g. [48, Ch. 10])

$$\tilde{D}_K(\xi) = \exp\left(\frac{1}{4\pi} \int_{-\pi}^\pi \frac{e^{it} + \xi}{e^{it} - \xi} \log K(t) \, dt \right), \qquad |\xi| < 1,$$

This $\tilde{D}_K$ is the outer function associated with $K^{1/2}$ normalized by $\tilde{D}(0) \geq 0$; in particular, $\tilde{D}_K$ is not zero in the unit disk, $\tilde{D}_K \in H^2$, and $K(\theta) = |\tilde{D}_K(e^{i\theta})|^2$ almost everywhere, where $\tilde{D}_K(e^{i\theta})$ denotes the nontangential boundary limit of $\tilde{D}_K$ at the point $e^{i\theta}$. If we set $K(t) = V(\cos t)$, then $D_V(z) = \tilde{D}_K(\xi)$, $\xi = z - \sqrt{z^2 - 1}$, where, as usual, we choose that branch of the square root that is positive for positive z. Hence, D_V is not zero in $C \setminus [-1, 1]$ and $V(x) = |D_V(x)|^2$ for almost every $x \in [-1, 1]$, where again the last quantity is a boundary limit.

The function Γ that we used above gives the argument of D_K on the lower part of the cut $C \setminus [-1, 1]$ (see [48, (12.1.7)] and note that $x = \cos \theta$, $0 \leq \theta \leq \pi$ on the lower part of the cut corresponds to the point $e^{i\theta}$ under the mapping $\xi = z - \sqrt{z^2 - 1}$):

(14.15) $D_V^2(x)/|D_V(x)|^2 = e^{i\Gamma_V(x)}.$

Theorem 14.3 *With the assumptions of Theorem 14.1 we have*

$$p_{n,n+k}(z) = (1 + o(1))\frac{1}{\sqrt{2\pi}}(z + \sqrt{z^2 - 1})^{n+k} (D_{V_n}(z))^{-1}$$

uniformly on compact subsets of $\overline{\mathbf{C}} \setminus [-1, 1]$, *where*

$$V_n(t) = w_n^{2n}(t)u^2(t)\sqrt{1 - t^2}.$$

We can put Theorem 14.3 into a somewhat different form (c.f. [43]). For simplicity let us assume that u is identically one (this can be attained in the most interesting cases by looking at $w_n u^{1/n}$ instead of w_n), in which case we will not need the Szegő function in our asymptotic formula.

Theorem 14.4 *With the assumptions of Theorem 14.1 we have that the polynomials* $p_{n,n+k}(z)$ *are asymptotically equal to*

$$\frac{1}{\sqrt{2\pi}}(z + \sqrt{z^2 - 1})^{k+1/2} \exp\left(nF_{w_n} - n\int \log\frac{1}{z - t}\, d\mu_{w_n}\right)(z^2 - 1)^{-1/4}$$

uniformly on compact subsets of $\overline{\mathbf{C}} \setminus [-1, 1]$, *where* μ_{w_n} *and* F_{w_n} *are the equilibrium measure and the equilibrium constants from Theorem A, Section 1.*

If u is not identically one, then the fixed multiplier

$$(D_{u^2}(z))^{-1}$$

also appears on the right.

This form gives via standard arguments the following asymptotics for the zeros of the orthogonal polynomials: with the assumptions of Theorem 14.1 let ν_n be the normalized counting measure on the zeros of the orthogonal polynomials $p_{n,n+k}$. Then

$$\lim_{n\to\infty} |\nu_n(I) - \mu_{w_n}(I)| = 0$$

uniformly in the intervals $I \subseteq [-1, 1]$. Of course, to conclude this we do not need the full force of the strong asymptotic formula in the preceding theorem, n-th root asymptotics would suffice (see e.g. [47, Chapter 3]).

That the two theorems Theorem 14.3 and 14.4 are equivalent can be easily seen from the fact, that the Szegő function associated with $\sqrt{1 - x^2}$ is

$$\left(\frac{\sqrt{z^2 - 1}}{z + \sqrt{z^2 - 1}}\right)^{1/2},$$

and, by Theorem A in the introduction the function

$$(z + \sqrt{z^2 - 1}) \exp\left(\int \log\frac{1}{z - t}\, d\mu_{w_n}(t) - F_{w_n}\right)$$

coincides with the outer function $D^2_{w_n}$ (on the domain $\overline{\mathbf{C}} \setminus [-1, 1]$) associated with w_n. Thus, it is enough to prove one of them, but before we do that let us utilize once more the preceding formula.

By what we have said about the relation of Γ_V and D_K in (14.15) it follows from the just metioned identity that

$$\Gamma_{w_n}(x) = \Im \left\{ \log(x + \sqrt{x^2 - 1}) + \int \log \frac{1}{x - t} v_{w_n}(t) \, dt \right\},$$

where $v_{w_n}(t)$ is the density of the equilibrium measure μ_{w_n}. Since the argument of $(z - t)^{-1}$ on the lower part of the cut of $\overline{\mathbf{C}} \setminus [-1, 1]$ is π for $t > \Re z$ and zero otherwise, it readily follows that

$$\Gamma_{w_n}(x) = \pi \int_x^1 v_{w_n}(t) \, dt - \arccos x,$$

and so we get the following variant of Theorem 14.2.

Theorem 14.5 *With the assumptions of Theorem 14.1 for $k = 0, \pm 1, \ldots$ the difference*

$$p_{n,n+k}(x) w_n^n(x) u(x)$$

$$-\sqrt{\frac{2}{\pi}} \frac{1}{\sqrt[4]{1 - x^2}} \cos \left(\left(k + \frac{1}{2} \right) \arccos x + n\pi \int_x^1 v_{w_n}(t) \, dt + \Gamma_u(x) - \frac{\pi}{4} \right)$$

tends to zero in $L^2[-1, 1]$.

Proof of Theorem 14.3. We use again (14.11). If we recall the properties of the Szegő function this can be written with $\xi = z - \sqrt{z^2 - 1}$, i.e. $z = \frac{1}{2}(\xi + 1/\xi)$ in the form

$$\int_{|\xi|=1} \left| p_{n,n+k}(z) \xi^{n+k} \tilde{D}_{K_n}(\xi) - p^*_{n,n+k}(z) \xi^{n+k} \tilde{D}_{K_n}(\xi) \right|^2 d|\xi| \leq 8\epsilon,$$

where

$$K_n(t) = \chi^2(\cos t) V_n(\cos t) = \chi^2(\cos t) w_n^{2n}(\cos t) u^2(\cos t)|\sin t|,$$

$n \geq n_\epsilon$, and, as before, $p^*_{n,k}$ denote the orthonormal polynomials with respect to W^*_n (see (14.5)). But here both functions under the integral sign belong to the Hardy space H^2, therefore we obtain from Cauchy's formula, that uniformly on compact subset of the unit disk $\{\xi \,|\, |\xi| < 1\}$ the difference

$$(14.16) \qquad p_{n,n+k}(z) \xi^{n+k} \tilde{D}_{K_n}(\xi) - p^*_{n,n+k}(z) \xi^{n+k} \tilde{D}_{K_n}(\xi)$$

tends to zero if $n \geq n_\epsilon$ and $\epsilon \to 0$.

By the choice of the functions H_n and h_n we have for

$$K^*_n(t) = 1/|H^2_{n-1}(\cos t)|^2$$

the identity

$$\tilde{D}_{K_n}(\xi)/\tilde{D}_{K_n^*}(\xi) = \tilde{D}_{(\chi h_n)(\cos)}(\xi)^2,$$

and since the $L^1_{2\pi}$ norm of $\log[\chi(\cos t)h_n(\cos t)]$ is as close to zero as we like by choosing $\epsilon > 0$ sufficiently small and n large (see Theorem 10.1) and of course selecting the function u^* in (14.4) appropriately, the ratio on the left of the preceding formula will be as close to 1 as we like. Hence it is enough to examine the behavior of

$$p_{n,n+k}^*(z)\xi^{n+k}\tilde{D}_{K_n^*}(\xi).$$

It is known that

$$p_{n,n+k}^*(z) = \frac{1}{\sqrt{2\pi}}\left(\xi^{n+k}\tilde{D}_{K_n^*}(1/\xi)^{-1} + \xi^{-n-k}\tilde{D}_{K_n^*}(\xi)^{-1}\right)$$

(see [48, Theorem 2.6, (2.6.2)] where this formula appears for $|z| = 1$ — actually in the form of the first displayed equation in the proof of Theorem 14.2 — which clearly implies the same formula for all z because both sides are polynomials in z), hence it is left to show that here the second term is the dominant one. But that is easy: the ratio of the first and second terms on the right has absolute value 1 on the unit circle. Since the Szegő function of a nonnegative trigonometric polynomial of degree l is a polynomial of degree l ([48, Theorem 1.2.2]), it follows that the ratio in question has a zero at the origin at least of order $2(n + k) - 2(n - i_n)$, where $n - i_n$ is the degree of H_{n-1}. Thus, by Schwarz's lemma the ratio of the first and second terms tends to zero uniformly on compact subsets of the unit disk as $n \to \infty$ because $i_n \to \infty$. This proves

$$p_{n,n+k}^*(z)\xi^{n+k}\tilde{D}_{K_n^*}(\xi) \to \frac{1}{\sqrt{2\pi}}.$$

From here we obtain the theorem if we use the aforementioned fact that the difference in (14.16) is as small as we like provided we choose ϵ small and n sufficiently large (and u^* in (14.4) appropriately), furthermore that with

$$U_n(t) = V_n(\cos t) = w_n^{2n}(\cos t)u^2(\cos t)|\sin t|$$

the ratio $\tilde{D}_{K_n}(\xi)/\tilde{D}_{U_n}(\xi)$ (which is nothing else than $\tilde{D}_{\chi(\cos t)}(\xi)^2$) can also be as close to 1 as we like (see the definition (14.4) of χ and recall that its geometric mean can be as small as we want).

                                                    ■

15   Freud weights revisited

Let $w_\alpha(x) = \exp(-\gamma_\alpha|x|^\alpha)$, $\alpha > 0$ be the Freud weights we considered in Section 3, and consider the orthonormal polynomials with respect to w_α^2:

$$p_n(w_\alpha; x) = \gamma_n(w_\alpha)x^n + \cdots.$$

We set for all n

$$w_n(x) = \exp(-\gamma_\alpha |x|^\alpha), \qquad x \in [-1, 1].$$

Note that w_n is defined on $[-1, 1]$.

Let $\rho_n = 1 + n^{-7/12}$ and

$$w_n^*(x) = w_\alpha(\rho_n x) = e^{-\gamma_\alpha \rho_n |x|^\alpha}, \qquad x \in [-1, 1].$$

It follows from the infite-finite range inequality (3.11) that

$$(15.1) \qquad \gamma_n(w_\alpha) n^{(n+1/2)/\alpha} \rho_n^{(n+1/2)/\alpha} = (1 + o(1)) \Big/ E_{n,2}((w_n^*)^n),$$

where $E_{n,2}$ is the extremal quantity from Section 13, (13.1) (see also the connection (14.2) in between this quantity and the leading coefficients of orthogonal polynomials). Here the support of the equilibrium measures associated with w_n^* is

$$[-\rho_n^{-1/\alpha}, \rho_n^{-1/\alpha}]$$

and not $[-1, 1]$ (see Section 3). But recall the discussion at the end of Section 9, where we proved that the estimates of Lemma 9.1 actually hold true on a larger range (see (9.26)), hence Corollary 10.4 is also true for symmetric weights $\{w_n^*\}$ which have support $S_{w_n^*} = [-\xi_n, \xi_n]$ with some ξ_n satisfying $1 - n^{-\tau} \le \xi_n \le 1$ for some $\tau > 0$, and which otherwise satisfy the assumptions of of Theorem 10.3 on $[0, \xi_n]$ rather than on $[0, 1]$. In particularly, this is true for the weights w_n^* we are considering now. But the proof of Theorem 13.1 was based on approximation like in Corollary 10.4, hence Theorem 13.1 is also true for such weights, in particular, for our w_n^*'s.

On applying Theorem 13.1 we immediately get

$$\gamma_n(w_\alpha) n^{(n+1/2)/\alpha} \rho_n^{(n+1/2)/\alpha} \sigma_2 2^{-n+1/2} G[w_n^*]^n = 1 + o(1).$$

Here $\sigma_2 = \sqrt{\pi/2}$ and, as we have already seen in Section 3.1, (3.19),

$$\gamma_\alpha \frac{1}{\pi} \int_{-1}^{1} \frac{|x|^\alpha}{\sqrt{1 - x^2}} \, dx = \frac{1}{\alpha},$$

which yields

$$G[w_n^*]^n \rho_n^{n/\alpha} = e^{-n/\alpha} \exp\Big((n/\alpha)((1 - \rho_n) + \log \rho_n)\Big) = (1 + o(1)) e^{-n/\alpha}.$$

Thus, we finally arrive at

$$(15.2) \qquad \lim_{n \to \infty} \gamma_n(w_\alpha) \pi^{1/2} 2^{-n} e^{-n/\alpha} n^{(n+1/2)/\alpha} = 1,$$

which is the extension of (3.3) to all $\alpha > 0$.

Let now

$$p_{n,k}^*(x) = n^{1/2\alpha} p_n(w_\alpha; n^{1/\alpha} x), \qquad x \in [-1, 1],$$

and let $p_{n,k}$ be the orthonormal polynomials associated with w_n^n, like in the preceding section. We apply the parallelogram law

$$\frac{1}{4}\int_{-1}^{1}(p_{n,n+k}-p_{n,n+k}^{*})^{2}w_{n}^{2n} + \int_{-1}^{1}\left(\frac{1}{2}(p_{n,n+k}+p_{n,n+k}^{*})\right)^{2}w_{n}^{2n}$$

$$=\frac{1}{2}\int_{-1}^{1}(p_{n,n+k})^{2}w_{n}^{2n}+\frac{1}{2}\int_{-1}^{1}(p_{n,n+k}^{*})^{2}w_{n}^{2n},$$

and observe that the first term on the right is $1/2$, the second one is at most $1/2$ since

$$\int_{-1}^{1}(p_{n,n+k}^{*})^{2}w_{n}^{2n}\leq\int_{-\infty}^{\infty}p_{n}(w_{\alpha};x)^{2}w_{\alpha}^{2}(x)dx = 1,$$

while the second term on the left is $1+o(1)$ because the ratio of the leading coefficients of $p_{n,n+k}^{*}$ and $p_{n,n+k}$ tend to 1 (see (15.2) and Theorem 13.1, and also recall (14.2)). Thus, it follows that

$$(15.3) \qquad\qquad \lim_{n\to\infty}\int_{-1}^{1}(p_{n,n+k}-p_{n,n+k}^{*})^{2}w_{n}^{2n}=0.$$

Using this relation instead of (14.13) everything that we have proven in the preceding section on w_n can be carried over to the Freud polynomials. For example it follows from Theorem 14.5 that for fixed $k=0,\pm1,\ldots$ the difference

$$n^{1/2\alpha}p_{n+k}(w_{\alpha};n^{1/\alpha}x)\exp(-n\gamma_{\alpha}|x|^{\alpha})$$

$$-\sqrt{\frac{2}{\pi}}\frac{1}{\sqrt[4]{1-x^{2}}}\cos\left(\left(k+\frac{1}{2}\right)\arccos x + n\pi\int_{x}^{1}v_{\alpha}(t)\,dt - \frac{\pi}{4}\right)$$

tends to zero in $L^{2}[-1,1]$, where v_{α} is the Ullman distribution (3.4).

We also mention that in [26] it was proved that pointwise asymptotics of this form are also valid (at least for $\alpha\geq3$, see also [43] for another proof which covers every $\alpha>1$), i.e. the above difference tends to zero uniformly on compact subsets of $(-1,1)$ not just in L^{2} norm.

On the other hand, it follows from Theorem 14.4 and

$$F_{w_{n}}-\int\log\frac{1}{z-t}\,d\mu_{w_{n}}=\log(z+\sqrt{z^{2}-1})+\int_{0}^{1}\frac{zt^{\alpha-1}}{\sqrt{z^{2}-t^{2}}}\,dt$$

(c.f. the computation in Section 3) that

$$n^{1/2\alpha}p_{n+k}(w_{\alpha};n^{1/\alpha}z)$$

is asymptotically equal to

$$\frac{1}{\sqrt{2\pi}}(z+\sqrt{z^{2}-1})^{n+k+1/2}\exp\left(n\int\frac{zt^{\alpha-1}}{\sqrt{z^{2}-t^{2}}}\,dt\right)(z^{2}-1)^{-1/4}$$

uniformly on compact subsets of $\overline{C} \setminus [-1, 1]$. In fact, using (15.3), the proof of Theorem 14.3 can be copied to yield the preceding asymptotic relation from that of the one in Theorem 14.4.

We could easily extend these asymptotic formulae for orthogonal polynomials with respect to a weight W^2 where W satisfies the conditions of Theorem 12.1. In fact, Theorem 10.3 can be applied in such case and otherwise the proof is just the same as before. Compare this with the results of [28] where similar results were proven under more restrictive conditions.

16   Multipoint Padé approximation

In this section we briefly discuss the problem of multipoint Padé approximation which is intimately connected to orthogonal polynomials with respect to varying weights. In fact, this area was the main motivation for A. A. Gonchar and G. Lopez [10] (see also [21]) for considering orthogonal polynomials with respect to varying weights and our discussion would not be complete if we did not touch this aspect of the theory.

Historically orthogonal polynomials originated from continued fractions, and one of the classical results in the analytic theory of continued fractions is Markov's theorem to be discussed briefly below.

A function of the form

$$(16.1) \qquad\qquad f(z) = c + \int \frac{d\mu(x)}{x - z}$$

is called a Markov function if μ is a positive measure with compact support $S(\mu) \subseteq \mathbf{R}$ i.e. Markov functions are Cauchy transforms of positive measures μ with compact support in $\mathbf{R}$. For functions of type (16.1) A. Markov [Ma] proved that the continued fraction development

$$(16.2) \qquad\qquad \cfrac{b_1}{z - a_1 + \cfrac{b_2}{z - a_2 + \cdots}}$$

of f at infinity converges locally uniformly in $\overline{C} \setminus I(S(\mu))$, where $I(S(\mu))$ is the smallest interval containing $S(\mu)$. In what follows we shall assume that the support of μ lies in $[-1, 1]$.

It is well known that the n-th convergent is the $[n - 1/n]$ Padé approximant to the function (1.1). Hence the convergents of (1.3) are rational interpolants with all interpolation points being identically infinity.

Gonchar and Lopez considered rational interpolants with more general systems of interpolation points. For every $n \in \mathbf{R}$ we select a set

$$A_n = \{x_{n,0}, \ldots, x_{n,2n}\}$$

of $2n + 1$ interpolation points from $\overline{C} \setminus I(S(\mu))$, which are symmetric onto the real axis. The points need not to be distinct. We set

$$(16.3) \qquad \omega_n(z) := \prod_{\substack{j=0 \\ x_{n,j} \neq \infty}}^{2n} (z - x_{n,j}).$$

The degree d_n of ω_n is equal to the number of finite points in A_n.

Denote by $\mathcal{R}_n$ the set of all rational functions with complex coefficients with numerator and denominator degree at most n. By $r_n = r_n(f, A_n, \cdot) \in \mathcal{R}_n$ we denote the rational function that interpolates the function f of type (16.1) in the $2n + 1$ points of the set $A_n = \{x_{n,0}, \ldots, x_{n,2n}\}$. If some of these points are identical, then the interpolation is understood in Hermite's sense. It is easy to see that this is equivalent to the assertion that the left-hand side of

$$\frac{f(z) - r_n(f, A_n; z)}{\omega_n(z)} = O(|z|^{-(2n+1)}) \quad \text{as} \quad |z| \to \infty,$$

is bounded at every finite point of A_n and at infinity it has the indicated behavior. We note that interpolation at infinity has not been excluded.

It can be shown (see [10] or [47, Lemma 6.1.2]) that there exists a unique rational interpolant

$$r_n(z) = r_n(f, A_n; z) = \frac{q_n(z)}{p_n(z)} \in \mathcal{R}_n$$

of the above type, and p_n satisfies the weighted orthogonality relation

$$\int p_n(x) x^k \frac{d\mu(x)}{\omega_n(x)} = 0 \quad \text{for} \quad k = 0, \ldots, n-1,$$

i.e. they are orthogonal polynomials of (exact) degree n with respect to the varying weights $\omega_n(x)^{-1} d\mu(x)$. Furthermore, the remainder term of the interpolant has the representation

$$(16.4) \qquad (f - r_n(f, A_n; \cdot))(z) = \frac{\omega_n(z)}{p_n^2(z)} \int \frac{p_n^2(x) d\mu(x)}{\omega_n(x)(x - z)}$$

for all $z \in \mathbf{C} \setminus [-1, 1]$. By homogeneity we can clearly assume that the p_n is the n'th orthonormal polynomial with respect to $\omega_n(x)^{-1} d\mu(x)$.

Suppose that $S(\mu) \subseteq [-1, 1]$. Now this is a typical situation when the assumptions in the results of Section 10 hold true, at least if the points of A_n are not too close to $[-1, 1]$. In fact, the function $1/|\omega_n(z)|$ can be written as

$$\frac{1}{|\omega_n(z)|} = e^{U^{\nu_n}(z)},$$

where ν_n is the measure that has mass 1 at every point of A_n. Thus, if we set

$$w_n(x) = \omega_n(x)^{1/(2n+1)},$$

then the equilibrium measure μ_{w_n} corresponding to w_n is nothing else than the balayage of $\nu_n/(2n+1)$ out of $\overline{\mathbf{C}} \setminus [-1,1]$ onto $[-1,1]$ plus

$$1 - d_n/(2n+1)$$

times the arcsine measure (equilibrium measure of $[-1,1]$). It is easy to verify that if the points of A_n stay away from $[-1,1]$, then the collections of all such measures has the property, that the corresponding densities are equicontinuous in compact subsets of $[-1,1]$ and they satisfy the conditions (9.1) and (9.2) with $\beta = -1/2$. Hence, the results of Section 14 can be applied provided the density u^2 of μ is in the Szegő class (see (13.5), and from the asymptotics there we can easily get strong asymptotics for the error away from $[-1,1]$. in view of (16.4) provided μ is absolutely continuous and its density u satisfies the Szegő condition (13.5). In fact, we know from Theorem 14.3 that $p_n(z)$ asymptotically equals

$$\frac{1}{\sqrt{2\pi}}(z + \sqrt{z^2-1})^{n+1/2}(z^2-1)^{-1/4}\left(D_{u^2}(z)\right)^{-1}$$

times the Szegő function (with respect to the domain $\overline{\mathbf{C}} \setminus [-1,1]$) of ω_n. Let d_n be the degree of ω_n. Then

$$\omega_n(z) = (z + \sqrt{z^2-1})^{d_n}\left((z - \sqrt{z^2-1})^{d_n}\omega_n(z)\right) = h_n(z)H_n(z),$$

and here the second factor $H_n(z)$ on the right belongs to the Hardy space $H^2(\overline{\mathbf{C}} \setminus [-1,1])$, and the square of the Szegő function associated with ω_n is just the outer function associated with H_n. Recalling that the ratio of H_n and that of its outer function is the Blaschke product (with respect to $\overline{\mathbf{C}} \setminus [-1,1]$) associated with the zeros of H_n (note that there is no singular part in H_n), it follows that what remains in the ratio in front of the integral in (16.4) is

$$2\pi(z + \sqrt{z^2-1})^{-1-2n+d_n}\sqrt{z^2-1}D_{u^2}(z)$$

times the Blaschke product associated with the zeros of ω_n. The integral itself converges to

$$\frac{1}{\pi}\int \frac{1}{t-z}\frac{1}{\sqrt{1-t^2}}\, dt = \frac{1}{\sqrt{z^2-1}}$$

by Theorem 14.1, so we have full description on how the remainder behaves away from $[-1,1]$:

$$(f - r_n(f, A_n; \cdot))(z) =$$
$$(1 + o(1))\pi D_{u^2}(z)(z + \sqrt{z^2-1})^{-1-2n+d_n}\prod_j \frac{\Phi(z) - \Phi(x_{n,j})}{1 - \Phi(z)\Phi(x_{n,j})},$$

where Φ is the canonical conformal map of the complement $\overline{\mathbf{C}} \setminus [-1,1]$ of $[-1,1]$ onto the exterior of the unit disk, and the product is taken for the zeros of ω_n, i.e. for the finite points in the system A_n.

The results of G. Lopez ([19]–[21]) give a different asymptotics that supersede the above one in the sense that the points in A_n can approach the interval $[-1, 1]$ so long as the sum $\sum(|(\Phi(x_{j,n})| - 1)$ tends to infinity.

It is worth mentioning the connection of the above asymptotic relation with the problem of minimizing the norms of Blaschke products on compact sets. In fact, suppose that V is a compact subset of $\overline{C} \setminus [-1, 1]$, and we want to construct good rational approximants to the Markov function f on V. By picking some sets A_n of $2n + 1$ points the above discussed multipoint Padé approximants are certain one of the candidates. The problem is how to choose optimally the interpolation points in A_n so that the approximation be as good as possible. In view of the asymptotic relation given for the error, our problem is to minimize the uniform norm of the Blaschke product in the error on V in the presence of the weight function

$$|D_{u^2}(z)|.$$

A result of Parfenov [41] is relevant here, which asserts that if V is an ellipse with foci at ± 1 (or the exterior of it), then the minimum behaves like $(r + \sqrt{r^2 - 1})^{-2n-1}$ times the geometric mean of the weight, where r is the sum of the half axes of V. It is plausible that the points of A_n can be chosen so that this asymptotic is attained. It is an interesting problem to investigate the same question for other, less symmetric V's. Another interesting problem is how far the so obtained bound for the above rational approximation to f on V is from the best one.

17   Concluding remarks

As we have seen, the method of Section 2 gives good approximation for logarithmic potentials by logarithms of reciprocals of polynomials provided the generating measure has continuous density. The method was sufficiently strong to settle the approximation problem for weighted polynomials $w^n P_n$ for a general class of weigths w and to considerably relax the conditions of [28] concerning approximation by weighted polynomials of the type $W(a_n \cdot) P_n(\cdot)$. It is possible that finitely many logarithmic type singularities in the density (these arise for example at the origin if one considers $w(x) = \exp(-|x|)$) can be handled by appropriately adjusting the correction polynomials (like $S_{n-[n/\lambda]}$ in Section 2) to have appropriate order of interpolation at these 'bad' points. The situation is much worse if the infinite singularity is not of logarithmic type. For example, if $w(x) = \exp(-c|x|^\alpha)$ with $0 < \alpha < 1$, then the density of the extremal measure has a singularity of the form $\sim t^{\alpha-1}$ at the origin, and indeed, we know that in this case approximation is not possible.

Internal zeros in the density function constitute another problem. We have seen in Example 4.5 that even a single zero may rule out the possibility of approximation in the sense of Theorem 4.2. On the other hand, Example 4.6 shows that in some cases approximation is possible even in the presence of an internal zero, and it seems to be a very delicate problem to clear the role of internal zeros on the approximation problem for given individual weights. The problem with internal zeros is that if the density function has a zero at x_0 in the interior of S_w, then in general x_0 will not belong to any S_{w^λ}, $\lambda > 1$ (c.f. the argument at the end of Section 6), and usually we need to apply the approximation technique to some w^λ instead of w in order to be able to handle the effect of the singularities in the density v that may appear around the endpoints.

A typical example of this kind of difficulty is encountered if we consider the weight $w(x) = e^{x^2}$ (note the positive coefficient in the exponent) considered on $\Sigma := [-1, 1]$. It can be shown that $S_w = [-1, 1]$ the density v of μ_w is given by

$$v(t) = \frac{2t^2}{\pi\sqrt{1 - t^2}},$$

which has a zero at the origin, and has a $(1 - x^2)^{-1/2}$ type singularity at ± 1 (see Section 11). If $\lambda > 1$, then S_{w^λ} will miss a neighborhood of 0. We do not know e.g. if for every function $f \in C[-1, 1]$ with $f(\pm 1) = 0$ there is a sequence of polynomials P_n of degree at most n such that

$$e^{nx^2}P_n(x) \to f(x)$$

uniformly on $[-1, 1]$.

Let us also mention that recently some efforts have been done to find a 'soft' approach to the approximation problem considered in this paper. In some restricted cases such an approach is possible, for example in [3] and [10] simple sign change counting was used to prove such theorems (this works for example if $w(x) = e^{-x^2}$). The paper [18] should also be mentioned that contains a construction for related "one point" polynomials.

$$* \quad * \quad *$$

Although I have not discussed the contents of this paper in details with D. S. Lubinsky and E. B. Saff, I would like to express my appreciation to them, because the present work was motivated by some of their results.

References

[1] N. L Akhiezer: On the weighted approximation of continuous functions by polynomials on the entire real axis. *AMS Transl., Ser. 2*, **22**(1962), 95–137.

[2] N. I. Achiezer: *Theory of Approximation*, (transl. by C. J. Hyman), Ungar, New York 1956.

[3] P. Borwein and E. B. Saff: On the denseness of weighted incomplete approximations, *Proceedings of the First US–Soviet Conference on Approx. Theory*, Tampa 1990, Springer–Verlag, (to appear).

[4] M. Brelot: Sur l'allure des fonctions harmoniques et sousharmoniques à la frontière, *Math. Nachr.*, **4** (1950–51), 17–36.

[5] Ch. J. de la Vallée–Poussin: Potentiel et problème généralisé de Dirichlet, *Math. Gazette, London*, **22**(1938), 17–36.

[6] G. Freud: On the coefficients in the recursion formulae of orthogonal polynomials, *Proc. Roy. Irish Acad. Sect. A*, **76**(1976), 1–6.

[7] B. Fuglede: Some properties of the Riesz charge associated with a δ-subharmonic function, (manuscipt)

[8] M. v. Golitschek: Approximation by incomplete polynomials, *J. Approx. Theory*, **28**(1980), 155–160.

[9] M. v. Golitschek, G. G. Lorentz and Y. Makovoz: Asymptotics of weighted polynomials, *Proceedings of the First US–Soviet Conference on Approx. Theory*, Tampa 1990, Springer–Verlag, (to appear).

[10] A. A. Gonchar and G. Lopez: On Markov's theorem for multipoint Padé approximants, *Mat. Sb.*, **105**(147)(1978), English transl.: *Math. USSR Sb.*, **34**(1978), 449–459.

[11] A. F. Grishin: Sets of regular growth of entire functions (Russian), *Teor. Funktsii, Funktsional. Anal. i Prilozhen. (Kharkov)*, **40**(1983), 36–47.

[12] L. L. Helms: *Introduction to Potential Theory*, Wiley–Interscience, New York 1969.

[13] X. He and X. Li: Uniform convergence of polynomials associated with varying weights, *Rocky Mountain J.*, **21**(1991), 281–300.

[14] K. G. Ivanov: E. B. Saff and V. Totik: Approximation by polynomials with locally geometric rates, *Proc. Amer. Math. Soc.*, **106**(1989), 153–161.

[15] K. G. Ivanov and V. Totik: Fast decreasing polynomials, *Constructive Approx.*, **6**(1990), 1–20.

[16] A. Knopfmacher, D. S. Lubinsky and P. Nevai: Freud's conjecture and approximation of reciprocals of weights by polynomials, *Constructive Approx.*, **4**(1988), 9–20.

[17] N. S. Landkof: *Foundations of Modern Potential Theory*, Grundlehren der Mathematischen Wissenschaften, **190**, Springer–Verlag, New York 1972.

[18] A. Levin and D. S. Lubinsky: Christoffel functions, orthogonal polynomials and Nevai's conjecture for Freud weights, *Constructive Approximation*, **8**(1992), 463–535.

[19] G. Lopez: Asymptotics of polynomials orthogonal with respect to varying measures, *Constructive Approximation*, **5**(1989), 199–219.

[20] G. G. Lopez: Szegő's theorem for polynomials orthogonal with respect to varying measures, *Orthogonal Polynomials and Their Applications*, Proceedings of Conf. Segovia, Spain, 1986, Eds. M. Alfaro et al., Lecture Notes in Mathematics, **1329**, Springer–Verlag, New York 1988, 256–260.

[21] G. G. Lopez: On the asymptotics of the ratio of orthogonal polynomials and the convergence of multipoint Padé approximants, *Mat. Sb.*, **128**(1985), 216–229. Engl. transl.: *Math. USSR Sbornik*, **56**(1987)

[22] G. G. Lopez and E. A. Rahmanov: Rational approximations, orthogonal polynomials and equilibrium distributions, *Orthogonal Polynomials and Their Applications*, Proceedings of Conf. Segovia, Spain, 1986, Eds. M. Alfaro et al., Lecture Notes in Mathematics, **1329**, Springer–Verlag, New York 1988, 125–156.

[23] G. G. Lorentz: Approximation by incomplete polynomials, *Padé and Rational Approximation: Theory and Applications*, Eds. E. B. Saff and R. S. Varga, Academic Press, New York 1977, 289–302.

[24] D. S. Lubinsky: Gaussian quadrature, weights on the whole real line, and even entire functions with nonnegative order derivatives. *J. Approx. Theory*, **46**(1986), 297–313.

[25] D. S. Lubinsky: Variations on a theme of Mhaskar, Rahmanov and Saff, or "sharp" weighted polynomials inequalities in $L_p(\mathbf{R})$, NRIMS Internal Report No. I575, Pretoria 1984.

[26] D. S. Lubinsky: *Strong Asymptotics for Erdős weights*, Pitman Lecture Notes, **202**, Longman–John Wiley & Sons, New York 1988.

[27] D. S. Lubinsky, H. N. Mhaskar and E. B. Saff: A proof of Freud's conjecture for exponential weights, *Constructive Approx.*, **4**(1988), 65–83.

[28] D. S. Lubinsky and E. B. Saff: *Strong Asymptotics for Extremal Polynomials Associated with Weights on* **R**, Lecture Notes in Mathematics **1305**, Springer–Verlag, New York 1988.

[29] D. S. Lubinsky and E. B. Saff: Uniform and mean approximation by certain weighted polynomials, with applications, *Constructive Approx.*, **4**(1988), 21–64.

[30] D. S. Lubinsky and V. Totik: Weighted polynomial approximation with Freud weights, *Constructive Approx.* (to appear)

[31] D. S. Lubinsky and V. Totik: How to discretize a logarithmic potential? *Acta Sci. Math. (Szeged)* (to appear)

[32] Al. Magnus: A proof of Freud's conjecture for exponential weights, *J. Approx. Theory*, **46**(1986), 65–99.

[33] A. Maté, P. Nevai and V. Totik: Strong and weak convergence of orthogonal polynomials on the unit circle, *Amer. J. Math.*, **109**(1987), 239–282.

[34] H. N. Mhaskar and E. B. Saff: Where does the sup norm of a weighted polynomial live? *Constructive Approx.*, **1**(1985), 71–91.

[35] H. N. Mhaskar and E. B. Saff: Weighted Analogues of Capacity, Transfinite Diameter and Chebyshev Constant, *Constructive Approx.*, **8**(1992), 105–124.

[36] H. N. Mhaskar and E. B. Saff: Extremal problems for polynomials with exponential weights, *Trans. Amer. Math. Soc.*, **285**(1984), 203–234.

[37] H. N. Mhaskar and E. B. Saff: A Weierstrass–type approximtion theorem for certain weighted polynomials, *Approximation Theory and Applications*, Ed. S. P. Singh, Pitman Publ. Ltd., 1985, 115–123.

[38] H. N. Mhaskar and E. B. Saff: Polynomials with Laguerre weights in L^p, *Rational Approximation and Interpolation*, Eds. P. R. Graves-Morris, E. B. Saff and R. S. Varga, Lecture Notes in Mathematics, **1105**, Springer-Verlag, Berlin 1984, 511–523.

[39] I. Muskhelishvili: *Singular Integral Equations*, P. Noordhoff, Groningen 1953.

[40] P. Nevai and V. Totik: Sharp Nikolskii inequalities with exponential weights, *Analysis Math.*, **13**(1987), 261–267.

[41] O. G. Parfenov: Widths of a class of analytic functions, *Math. USSR Sbornik*, **45**(1983), 283–289.

[42] E. A. Rahmanov: On asymptotic properties of polynomials orthogonal on the real axis, *Mat. Sb.*, **119**(161)(1982), 163–203. English transl.: *Math. USSR Sb.*, **47**(1984), 155–193.

[43] E. A. Rahmanov: Strong approximation on orthogonal polynomials associated with exponential weights on **R**, (mannuscript)

[44] E. B. Saff: Incomplete and orthogonal polynomials, *Approximation Theory IV*, Eds. C. K. Chui, L. L. Schumaker and J. D. Ward, Academic Press, New York 1983, 219–256.

[45] E. B. Saff, J. L Ullman and R. S. Varga: Incomplete polynomials: an electrostatic approach, *Approximation Theory IV*, Eds. C. K. Chui, L. L. Schumaker and J. D. Ward, Academic Press, New York 1983, 769–782.

[46] E. B. Saff and R. S. Varga: On incomplete polynomials, *Numerische Metoden der Approximationstheory*, Eds. L. Kollatz, G. Meinardus, H. Werner, ISNM **42**, Birkhäuser–Verlag, Basel 1978, 281–298.

[47] H. Stahl and V. Totik: *General Orthogonal Polynomials*, Encyclopedia of Mathematics, **43**, Cambridge University Press, New York 1992.

[48] G. Szegő: *Orthogonal Polynomials*, Amer. Math. Soc. Colloq. Publ. **23**, Amer. Math. Soc., Providence RI 1975.

[49] V. Totik: Fast decreasing polynomials via potentials, *Journal d'Analyse Mathématique*, (to appear)

[50] V. Totik and J. L. Ullman: Local asymptotic distribution of zeros of orthogonal polynomials, *Trans. Amer. Math. Soc.* (to appear)

[51] M. Tsuji: *Potential Theory in Modern Function Theory*, Maruzen, Tokyo 1959.

[52] J. L. Walsh: *Interpolation and Approximation by Rational Functions in the Complex Domain*, Amer. Math. Soc. Colloq. Publ., Amer. Math. Soc., Providence RI 1935 (Fifth edition, 1969).

Index